Introduction

In this world, we all see spiritual and physical life from our own frame of reference; so, our views are all different, at least to a certain degree. In this fallen world, we experience fear, peace, anger, suffering, joy, anguish, loss, and desperation. And our spiritual interpretations of it all, are colored by our core desires that lead us to our chosen lively forms of love. This makes our soul's experiential awareness, personal and intimate to us. But if we were all just soulless physical systems, as material science says, how could, and why would we desire and search for spiritual awareness, or faithful eternal love. How could, and why would any purely physical thing, ever know, desire, or seek things like eternal love and life? How could any-*thing*, being purely physical, seek out an eternal purposed Godly understanding? How could a purely physical creature ever truly love? How could any temporary purely physical system and process, ever desire a higher non-physical eternal spiritual life? Why would a form of pure physical complexity, ever have a desire for truth? Why do I seek a perfect eternal love, truth, and life under GOD? Because I am an eternal soul; I am *not* a purely physical creature. My soul's knowing belief, arises from my core desired need for eternal love and truth. How can any soulless physical aspect, need to know, have, and share love and all other mysterious non-physical wonders? I tell you the purest truth, the physical side of reality knows and desires nothing! Here in this world, we never see, hear, or know, the exact same thing; but we can become of one mind in going to, and knowing GOD?

Author, Lawrence Michael Stack: Theorist, Philosopher, Metaphysician and Theist

Friends call me Michael

ISBN 978-0-578-94754-9
LCCN 2021913495

Other books by this author…

1. The Theory of Created Conscious Perception
2. The Eternal Perspective
3. Live to know the GOD of Prophecy

Lower-case italics, implies mild emphasis added.

3 ALL CONSCIOUS FORMULATIONS

This vast complex forcefully aligned creation, is touched and held between the outstretched hands of FATHER for its continuity, and to keep it a physical part of him. FATHER'S desirous *forceful* spoken words formed it all, and it is only his desires that still sustain the physical half of reality in all its complex dynamic ways. FATHER'S planning and loving contemplation, prior to his spoken firmer external creations, were the *conscious formulations* of his purposed desires and ways. And FATHER'S subsequent gift of all souls, fulfilled his need and hope to love and be loved. And now for us all, even to GOD, our final outcome is wrapped in a mysterious cloud of free-will uncertainty. This third book on the grand subject of GOD and Creation, focuses more on the desired forethought and plan seen in FATHER'S and our creations. As I gaze out into the increasing horrid hateful conscious condition of this fallen reality, seeing it through GOD'S eternal perspective. I see every horror is attributed to the imperfect ways of fallen angels and mankind, and not once attributed to the Righteous Holiness of GOD in any way or part. It was then and now, the gifted and created who rebel against the perfect eternal ways of life FATHER created in his loving hope that it would eternally please all orders of lesser gifted souls. But we see FATHER'S created ways are not enough for most souls; yet they *are* enough, and perfect to the fewer. All observed and felt pain and loss, comes to exist in creation, by individual contrary desirous formulas, projected to now fill this sphere of reality with growing conscious and physical confusion, deception, division, hell, and death. Most souls have no desire to see or claim their conscious forceful contributions to all disorder. Most souls, being individuals, want only to temporarily exist, and sense the temporary pleasures of life's firm experiential surface. The lost only believe in what their physical eyes see; as its eternal spiritual reason and purpose are rarely considered. In most souls, their driving imperfect desires, and incomplete physical information, make it hard for them to see or know otherwise. In psychology it is said, only a few will commit suicide, yet most souls self-destruct. Meaning they gamble, smoke, drink, cuss, and are unfaithful; destroying their life's trusted relationships of love for and from others. And once a soul truly denies the existence of a Conscious Creator, they gamble their soul and become a Materialist. And seeing the expanding state of reality, a Materialist becomes a Physical Reductionist. But true Judeo Christians are also reductionist of a kind, as we reverse the fallen unfolding events of what led to what we see now, back too, and stop at the finished initial perfect *stable*, unstretched dynamic creation of GOD. But a Materialist *must* continue with a reductionist method, until reaching a singularity of unimaginable compressed energy. But there, their understanding and belief fall apart, as they can't mathematically reconcile any existence of force that could cause *and* hold such energy in an initial state of a singularity, prior to a Big Bang. First, they must stop the deception, and admit a true singularity can never exist as *two* balanced waring qualities; being an unimaginable explosive energy, held by an unimaginable attractive stabilizing force. The term singularity, means one pure quality, period. So, their belief becomes a fantasy beyond any statistical chance of ever being true. We explain FATHER'S vast spoken forth *external* initial spherical aether in prior books, formed of the pure countless physical singularities that form FATHER'S body; which we stand by. Yet some ask why we stop there in a reductionist definition of the body of FATHER GOD, to form the physical side of his spoken forth *external* creations?

If one questions whether singular spherical physical points, or one vast fluid spherical state of one pure physical quality, represents the substantive body of FATHER GOD. Both work; so, it matters not if FATHER'S physical body initially exists as a vast perfect physical fluid quality he forcefully divided and spoke forth as quantum physical singularities to form his *external* aether. We see only two forceful aspects in all created reality: one being the creative, Godly, and eternal, and all that is temporary destructive and rebellious to the same. In eternal soul Dualism, we see zero distinctions between the terms, perfect, and eternal. For a thing to remain eternal, it must be perfect. It must have no internal weakness, faults, or waring aspects. It must by its purity, be impervious to change by any external influence. So, if any basic aspects of your belief, changes as your life unfolds, then as of yet, you hold no eternal understanding or belief. Life is our book and our mind of ordered moral and physical imperatives, choices, and results. And all *eternal* spiritual and physical imperatives, choices, and reasons, are established and only sustained within FATHER'S created orders of life. All that FATHER desires and touches, he does for an eternal cause, reason, purpose, and affect. Our first spiritual and physical imperative and choice for eternal life, rests in our faithful responsible perpetuation of Godly love. And in being under GOD; faithful means eternal. If *your* life does not meet this definition, your life remains in its certain ways, outside of GOD'S eternal life. In soul Dualism, souls project creative forceful patterned control to create their own particle-wave based perceptions. And a soul's perfect immense, or minute *physical* creation, can only eternally exist, by and for its creator. All free-will inhabitants of a FATHER'S creation, must exert perfect *physical* control, or it is destroyed from within by contrary forces. Upon creating anything physical, its creator cannot change their desires for that creation, without altering said creation; and by the *conscious* laws of created reality, physical perfection cannot self-alter. Yes, within a vast perfect physical creation, its Creator in search of eternal love may divide out from himself, lesser forceful conscious agents having free-will, to dwell within his perfect physical creation. But by doing this, we see both love and a rebellion of contrary formulated spiritual and physical ways, can arise within that created perfection. Free-will is truly an absolute prerequisite for love to exist. We see that JESUS, HOLYSPIRIT, and many, by their free-will, have desired chosen believed and lived their entire existence in loving responsible faithful obedience to the initial truth of reality, that FATHER is the source of all that is good. JESUS has chosen in his every moment and experience since his gifted creation, to follow FATHER in perfection. Souls not seeing, knowing, or caring, project their immoral physical disorders into FATHER'S perfect creation; but only at his temporarily allowing such rebellious acts that bring a soul's separation into hell and death. FATHER'S Creation still serves his eternal purpose of finding eternal love, despite all the desired chosen rebellion that now fills it. FATHER'S creation is now steeped in false conscious forceful formulations of endless temporary beliefs. Materialists say the physical realm is all there is, as they try to understand it. While refusing to see that all the physical information they study is what remains of FATHER'S consciously formulated initial complexity. Information can be far smaller than any pattern in matter or energy that mankind can perceive. Mankind's perceptive ability is so limited, it reveals their inarguable inability to ever sense or understand reality's fulness. Science struggles to see, equate, and define physical reality in a way that works universally; but they can't, because they won't. All creative and destructive forces, reside in quantum bonding, radiation, motion, momentum, and oscillation, yet mankind can't sense that smallest fastest information that underlies and defines the greater realities. They can't even know how small or large the part is, they can sense and study. For mankind, physical reality is like a very thick book, that has all but a few pages in the middle, written in words to large and small for them to see. Mankind can read and study those fewer pages, yet those pages are insufficient, so they can never correctly interpret or understand those pages, without GOD revealing the rest. Materialist speculates for themselves, a miracle fantasy beginning, and a fantasy end for it all. But all their theories to date have failed, forcing them to support great absurdities that are literally impossible to test; which again proves their perspective cannot grasp realities fulness. They chose a dead-end science with no way back. All the evidence says our desired formulated understanding, creates our own internal belief about reality. So, why do materialist say all reliable information originates on the physical side, when even they clearly see the opposite? Our every complex desire, creates its forceful reflection of a particle-wave-based perception; with zero evidence or reason to see this order being created in any purely physical way. Our language and belief of accepted understanding, define how we sense. And the eternal perspective, seeks to projects all forceful desires through the one shared pure belief of GOD; while temporary perspectives, project every desire in their individual contrary ways. We find it is a true constant, that our soul's awareness resides *up* in the quantum realm; yet only being adjoined to the firm reality below, enables our wakeful interactions. So, by false beliefs, reality is misinterpreted and disordered by most souls. The ability to understand reality is lost to souls who believe all information moves in physically determined ways; despite that being disproven by quantum mechanics and entanglement. Yet this disproving is denied, as they seek a unified physical theory that does not exist. And a hundred+ years after the quantum realm became known, the uses and tech advances brought by quantum mechanics, blinds them as they grow evermore confused, grasping at fruitless insane suppositions and physical mysticisms and miracles. Please see that all the firmer physical complexities of reality are layered patterned formulas, consciously forced down, from their higher quantum realm foundation. The desirous *eternal* free-willed half of creative awareness that is our soul, has been gifted; but in Dualism, it must be eternally enabled by our reality's physical half, requiring FATHER'S creative formula. Fallen mankind, can't even know or define what is truly slow or fast, or what is time and eternity, or what is small or large, or what they can sense or can't sense. So, they can never honestly say beyond all possible doubt, they know one defined or measured part of reality. There is no aspect of time in which to go forward or backward; there is only every spherical soul's ongoing true or false complex beliefs, raised in every spiritual and physical experiential moment of their existing. Mankind's adjoined fallen sensing brain/body, is far too slow, and their perspective is far too small to sense or know reality on their own. And by those same limitations, they are blinded to that same fact. A soul's fallen physical perspective is limited in sensing, and subject to false beliefs that arise in seeing the disordered effects of entropy, fission, and a belief in time. And believing these were all there as a part of a false initial cause and effect, not seeing the original order of a perfect conscious complex design. In all abstract ways and understanding, time and eternity *share* the aspects of forced motion and sensed experience. And even fallen mankind knows that if anything moves or oscillates fast or slow enough, it will disappear from their fallen perceptive ability, as it moves into a higher eternal realm perspective. Each fallen soul's limitations rest in that they can't sense any aspects of reality that are too slow or small, or too fast or too large and too ethereal. So, what does disappear mean to a materialist; it means something they once sensed, they no longer sense in a comprehendible way. But they can't say it no longer exist, or was annihilated. Even with all future technical ways and means, fallen mankind can *never* sense, measure, or know, the size and speeds involved in this entire conscious-physical reality, because their perspective cannot not allow or enable such understandings. Has fallen mankind *even* accurately sensed and measured the six percent sliver within their perceptive reach? No. And if they ever do, they must still interpret it improperly. From their fallen too small and slow perspective, all they can do is sense, measure, and compare the statistical contrast within the tiny part they do sense. They can only know that among all the aspects they sense, some are bigger or smaller than others, some go faster or slower, some are firmer and some are not, some are more energetic while some are less so. But they can *never* know what exists beyond the ability to perceive. They have no true foundation to form any first immutable knowledge. So, their every physical question, and every physical answer, remain wholly inadequate. Fallen mankind can never truly measure any time-frame, or truly know the history of this creation; yet they endlessly theorize, believe, and quibble among their too small too slow individual interpretations of the tiny bit they think they know. A Materialist is like a child pondering rocks or blades of grass, as their body get old and dies. They can't see that after falling so far, mankind can only see a 4D reality, minus our eternal souls. They do not see how they go from one form of being wrong, to another, and another, without end. They hold to a false confused belief that they truly understand certain parts of reality. Reality can never be known through their distorted fallen logic and mysticisms, or by all the physical miracles they believe in. About twelve thousand years ago, every point and pattern of FATHER'S creative physical information, was spoken forth in a forceful consonance, to form everything in every direction minus all disorder or fission. Now what can we see? Can we see a pen, *appearing* perfectly still on a desk? Yes. But can we see a pen that is literally entirely perfectly still on a desk? No. All forcefully sustained physical reality, must oscillate and emit photons according to “alpha” enabling its radiation, and only then can we see that pen. Every created quantum aspect in a pen is oscillating, absorbing, and releasing the photons we sense, that only represent that pen. If every atom, particle, and singularity in a pen, lost all sustaining force, yet held together being perfectly still, which is impossible; it must be invisible to us. Pure unenergized physical singularities exist within FATHER, but are not a part of any external reality. FATHER’S desired patterned force, formed the energized designs of motion and complex reality that enable our sensed awareness of that reality. FATHER can’t desire and speak into his aether, without that desired pattern becoming real. Now, can any soul, even FATHER, ever see non-existence? No. And FATHER alone sees the sphere of his aether. Even seeing the dark vacuum of space, is seeing a created reality within that aether mankind calls the Cosmological Constant. We could not see the dark coldness of space, if it was not physically created and energized. Force creates all particles and waves, and each higher wave frequency equals greater energy and light. When souls see darkness, why do they say they see nothing when darkness is a real something, giving off just enough light for us to see it as darkness. But again, we never see any actual aspect of reality, including darkness; we see only our created representations of it all. Does this sound like circular reasoning? I hope so, as all true eternal logic *is* circular. Yes, the material sciences strongly disagree with circular logic, as they falsely believe in linear logic. If we could see the eternal logic of GOD being formed, we see it forms in an advancing circling motion, considering every possible aspect, till they acquire a perfect eternal spherical understanding. Only after a starting point of understanding, circles and spirals to become spherical, and at the last, finds and joins its first to its last; only then do you have one perfect eternal logical point of understanding, or physicality. And a vast immense sphere, formed and filled with all countless *relative* spherical points of perfected logic, forms the perfect eternal mind and belief of GOD. Now if any point of any imagined logic, circles endlessly and its first forever misses rejoining its beginning, as we see in fallen mankind's broken understanding, it must spiral downward in an endless linear trap of false ever-changing beliefs. Truth and wisdom are set and eternal; but lies are not. Denying FATHER'S spherical model of reality, and accepting anything linear, drags one down. How can true logic keep running away in a linear fashion, luring souls deeper into confusion? Any logic that does this, is an endless delusional search that tries to unify a false linear beginning to the false end of the same, as they forever run in opposite directions you can never see. The true answer to any true question, is sustained by its omniscient spherical information. And this only occurs among GOD'S family, who share this desire and knowledge, each to our own degree. We must desire and seek the Godly bonding spherical eternal loving ways of life itself; or join the fallen to exist in the linear dying ways of our individual physical desires. All who resist GOD, can't see what they unrightfully temporarily steal; nor do they comprehend all they destroy; or know what they forever lose. FATHER'S created ways of eternal love and life are unacceptable to the rebel, by being spherical, pure, complete, and final; not being an endless linear ever-changing self-ruled individualized process. Love and eternal peace among many, can only exist in its perfect ways and environment, protected and secured within GOD'S vast perfect sphere; we then being perfect tiny spheres within and part of GOD'S perfect greater sphere. It is not about our love falling uncontrolled, inward in our hearts as individuals, it is about our love and choices coming back around to what is *eternally* good for all, and not for any one soul. Do you not see how CHRIST has lived his entire spherical moment in his obedient love for FATHER; he has been our true Holy example for all souls since his gifted creation. If you resist a loving sacrificial life towards GOD, you cannot deny your desires are born of selfish individual physical pride, envy, and covetous. Do not deny this self-evident cause, for there is no other cause or place, from which any spiritual aversion to GOD'S shared eternal purity may arise. So, all ungodly resistance equals disorder, brought only by individualized *conscious* determinism. FATHER'S projected information, like ours, is a generated forceful aspect. And to allow choice to run its course, he will not at this moment, stop contrary projections of force. All force is of one same force in all its strengths and choices. So, FATHER'S creative patterns of force, can be reenforced by our tiny force wave convergence, or diminished by mankind's combined wave interference that now relentlessly tear at his creation. The created truth of GOD is the greatest of things, which must be learned step by step in GOD'S designed way of our being led to know. Most souls think free-will should allow them to change in an instant at expressing a surface thought of doing so; but free-will is not a set of sudden switches on our soul's surface. It is in our built up chosen deep core desires, where we see a process and effort and cost are required for any real free-will change. There is our most weak free-will desire, and our most strong profound free-will desire. Yes, we may change our mind on an ice cream cone from chocolate to strawberry in an instant. But to truly turn to return to GOD, means our life-long most profound free-will desires of belief we have built deep within us, are a mountain we must tear down and rebuild. I know this, because this is what *I* have found. Our free-will strength and weakness to change our core desires, are felt and seen from our surface to our core. To change our core desires, requires a sustainable conscious formula based upon a pure Godly spiritual understanding; not just a few new physical habits. Our strong worldly desires are our built-up conscious constructs; and in basic ways, our core desires are our conscious directives of powerful influential currents coursing throughout our soul and body. When too strong, these pull souls over a precipice of self-destruction. Now though not every slight physical indulgence of pleasure is seen as a firm act of sin, they weaken us towards sin. If we are to change our deep profound free-will desires, we must sustain a forceful effort of crucifying all desires of the flesh. We know that as we build the new, we must tear down the old; or we will *always* face our same waring desires. In making sustained life-altering directional changes, we can't build the new and good beside the old and evil. And we must learn and know why a certain desire is evil, in order to truly tear it down. In FATHER'S design, before any rebellion, a soul's learning was a painless pure and true process. But now for each fallen soul in a fallen creation, constantly assaulted by the fallen desires and beliefs of others, it is most often a hard choice and process. So, we see the enduring strength of our desire to change is a most crucial factor. We can only find GOD'S basic truths, through our strength to lovingly believe their every word without question. We hold and strengthen all we search and hope for, and have been graciously given. In coming out of rebellious curiosity, our firm foundation of GOD'S truth is gained through suffering, loss, and total surrender. Pure Dualism vs Individualism, show us there can never be one enduring belief among many lesser souls, unless it comes from GOD. And according to souls who poll and analyze this world with biased probative questions, they say over two-thirds of all souls believe there is *a god*; yet most say they do not know GOD intimately. To truly believe and love GOD, we must know them intimately. Do you see how this world believes in their many imagined gods, which are not gods. If this world does not know the one true GOD, it is godless, no matter how many false gods they believe in. The one true GOD, like all imagined gods, does not come to exist, at the conjuring's of people's belief. When souls search for, find, and know the one true GOD through the spiritual and scientific reasons Scripture reveals through HOLYSPIRIT, they see we are *called* by Scripture to learn, know, love, and fully emulate the life of CHRIST. By FATHER'S commands, we are all called to intimately and correctly know CHRIST'S perfect attributes, so we may live in his spiritual semblance, above all individualism, and this world. What do *you* sense GOD teaching throughout Scripture? Do you sense one specific common shared way of life in mind for all Children? Or has GOD said for each soul to interpret and define GOD'S words for themselves through their covetous desires? I hope you each will please honestly ask yourself if you hold worldly desires and weakness? And if so, can you truly and correctly interpret Scripture on your own? Or, do you know your infirmities must cause you to misinterpret, and over or under emphasize many spiritual and physical things? Do you trust yourself or another person to interpret GOD'S meaning in each word of Scripture? Or do you know only HOLYSPIRIT can teach you this truth, as Scripture clearly says? Does this cause you to question your faith that HOLYSPIRIT is truly there, or if will he come and dwell within you in your current spiritual state? Surely, there is only one shared Holy Righteous belief and way of life for GOD'S children. We see familial faithful responsibility is a founding Principle for all bonding of Sentient-Beings. So, what does a Godly spiritual corporeal life look like? Well, in Scripture, we see FATHER, JESUS, HOLYSPIRIT, the Holy Angels, the Saints, Prophets, and Martyrs as our examples. And we see in every physically embodied example, that JESUS, the Saints, Prophets, and Martyrs, all spiritually knew physical death meant going to be with FATHER. And CHRIST said, in our being of him in *his* fulness, we are then spiritually and physically as the Holy Angels. We are called to transform ourselves here and now, and not wait or expect GOD will over-ride our fallen free-will desires so we can enter Heaven; that will never happen. Our souls need to align in one freely-chosen shared state of

Spiritual-Metamorphous to awake, hear, see, understand, and experience, every day in growth. Our command from CHRIST is to here and now become a Spirit, or Angel of GOD; to remain strong in CHRIST in a process of eternal love and wisdom. Our learning of, and becoming a Godly quality and quantity is perfected, as we each endure this other-worldly spiritual metaprocess. A glorious love is felt *from* GOD above to light our path, and this aspirational eternal path is right here before us all; yet it is only seen, traveled, understood, cherished, and loved, in a soul's deep desire to do so. Can and do all go to GOD as called? Or do many say they will wait where they are in a belief GOD will come for them at the last, refusing to leave their individual ways and desires? Scripture clearly states that many are called, but few are chosen. And why does Scripture also say that from among all the Righteous, GOD'S Children are scarcely saved? Why does Scripture say those found with CHRIST at his return are all called, *and* chosen, *and* faithful? Do you exhibit these *three* required spiritual attributes? Has GOD *called* you? Has GOD also *chosen* you? Not have you chosen GOD. And are you in an enduring manner, eternally *faithful* to GOD'S WORD? Heaven *is* a place, but it is far less a place, and far more a shared eternal responsible fervent desire, deep within each soul's eternal faithful loving heart. Can you grasp what an eternal faithful aspiration for GOD is? Don't think GOD has asked you or I for those three qualifiers, and set them beyond our abilities. Those three attributes are in *our* giving part, being *our* loving act from *our* side of love's eternal equation. GOD can love and give us life in abundance; but GOD cannot instill the spiritual quality of eternal love for them, within anyone. This is the choice and work of eternal love we make on our part; and it can never become real in desperate self-preservation at one's last moment. Do we tell family and friends there will be time for us to love them dearly later on our death bed? Love is not a switch we flip, eternal love is an eternal commitment and effort of enduring desire to become one in eternal care for the good of GOD'S Kingdom. And it is to us, to desire, choose, *intimately* learn of, and love GOD; or not love them enough, or not at all. Do any think GOD should have eternal love for us, and at the last moment, somehow instill in us our eternal true love for them? Must GOD provide all love from both sides? Has FATHER not created chance, choice, and gifted our eternal free-will to choose and strengthen our own desires, expressly to be loved by all who choose him? Please do not live in any partial denial of GOD'S truth. Our eternal conscious soul, our final physical platform, and our eternal estate, rest solely upon our final belief and understanding of love, being our formed desired spiritual desired awareness towards all great eternal things. Most of mankind, and most leaders of this world, tear at all who lovingly obey GOD. We wish for JESUS to come soon, bringing an end to every new evil; yet we also need our expanding spherical moment to strengthen and prepare ourselves, and add souls to GOD'S Kingdom. A coming last foretold moment, rests solely with FATHER; yet it comes not by his arbitrary choice. It comes when FATHER sees that only individual zombie reprobates remain, seeing no hope for a single new eternal love, making any further search futile. Imagine the time of the great flood, minus those eight souls who still loved GOD; imagine not one soul left upon the face of the earth who might hear and repent. All are turned firmly to their darkness of desires and false beliefs, being their individual spiritual hell and death. The purposed temporary allowance of choice and evil, will have run its full course of diminishing returns. And the end of *this* third dispensation of grace, will close. Yes, that will be the end of CHRIST'S grace and suffering to pay for the sins of others; that Grace, there and then ceases. And for CHRIST'S Millennial Reign on earth, in Ezekiel we read the animal sacrifice returns, and CHRIST will rule with a rod of iron. Of all the souls *called*, many cannot want the fullest understanding and beliefs of GOD'S eternal perspective. We cannot let ourselves or relationships with this world, be our focus; think always of GOD. We must seek our perfect humble place, if we are to hear and accomplish our eternal created purpose. Trust, pray, and know, that GOD'S perfect-will can never be done by souls living in perpetual sin, filled with individual worldly understandings and beliefs. In this world, most all things including belief, now come in endless desired versions, styles, and every changing form imagined; and that is all they are is imagined. The eternal perspective sees all imagined realities to be individual delineations of divisive expression, being not commanded, sustained, or created by GOD. So, what is the truth? Is it of GOD, or as this world says, something to be taken by each soul and bent to conform to their individual desire along a spectrum. All beliefs come from one of two perspectives: one is the first to last word of GOD; the other is through *endless* confused changing individual words. Can there ever be more than one truth, seeing GOD does exist? Souls can rebel and create beliefs from individual perspectives; but no one can make their beliefs rule over reality, in any sense. Souls who are not aligned, yet intrigued, wonder what would this highest form of spiritual love feel like? Does it have *any* physical aspects and sensations? Being in GOD, I do experience, know, and understand, how and why eternal love exists; and yes, in our being bicameral entities, eternal love has its expressions running through our physical side, as in how we experience all emotional things. In strong Godly moments, I spiritually *and* physically sense HOLYSPIRIT'S assurance stirring and moving within my soul *and* this body. We know our soul's senses and emotions have flow and feed-back, that circulate through the Dual aspects of our body and soul. If asked to *fully* describe this higher feeling of love, I am afraid that cannot be done. This quale, as with many, is experiential and impossible to fully convey. I can say it is the truest form of peace I have ever felt, and often it is so strong, it proves to me there are two foundations and forms of love. And GOD'S love being eternal, is so much stronger and has such an assuredness, it feels other-worldly. Now other-worldly is a term I use to describe GOD'S every aspect. GOD'S love has a forceful cleansing aspect, yet our souls are surrounded in this world with evil considerations, trying to dominate the awareness of our moment. All we each know and believe, fills our sphere; yet our current moment of strong awareness, rests at the surface of our soul, interacting with all realities beyond our soul. When we project love from our surface, to another soul's surface in this world, we see that other forces fight to dominate each soul's surface; even the local ambient environment is a factor. Many forceful considerations try to occupy each soul's surface, and crowd out the flow of love. Love is often felt as a blurred, thinly vailed expression in this world; but in growth, eternal love can be felt to dominate. When we are finally with GOD in their Eternal Kingdom of provision, considerations not of love, will barely approach the under surface of our awareness. We will exist in a loving elation and euphoria, that short of experience, is impossible to know or describe. To suddenly be before FATHER, and hear that he knows and loves you, and you have no worry or fear of new loss or pain. Then no negative concerns can exist, just pure positives of love. This is no sufficient description, as qualia can never be described, it must be experienced.

Even in Heaven we will never be able to adequately describe our love to each other. It can be hard to accept such a shared love can exist that can only be known if we no longer love this world, or this body, and we see our soul as a treasured eternal reality. This highest love is eternally sustained by GOD fulfilling our every shared need and alleviating every other concern. There will be no worry or questions about provision, belief, ability, race, or gender; such questions will not exist. A soul is not male or female, black, brown, or white, a soul is not tall or short; we are all measured distinct spherical souls, meant to exist, connect, and overlay in love, belief, and truth, like three-hundred-million distinct Russian nesting dolls. We are currently purely good, or mixed evil and good polarities of forceful desire and belief in our self-awareness. Please consider what was truly set at stake, for us, in how CHRIST in eternal love, temporarily put off his eternal body of GOD and put at risk his soul to come to this horrid world to harshly exist in the fallen flesh of mankind. Yet he *never* sinned once as he suffered among great all this imperfection and vile for 33 years, and then suffered more than anyone has ever suffered as he died and went into the pit for three days and nights to preach his loving truth. Then

WHICH YE HAVE HEARD FROM THE BEGINNING SHALL REMAIN IN YOU, YE ALSO SHALL CONTINUE IN THE SON, AND IN THE FATHER. AND THIS IS THE PROMISE THAT HE HATH PROMISED US, EVEN ETERNAL LIFE. THESE THINGS HAVE I WRITTEN UNTO YOU CONCERNING THEM THAT SEDUCE YOU. BUT THE ANOINTING WHICH YE HAVE RECEIVED OF HIM ABIDETH IN YOU, AND YE NEED NOT THAT ANY MAN TEACH YOU: BUT AS THE SAME ANOINTING TEACHETH YOU OF ALL THINGS, AND IS TRUTH, AND IS NO LIE, AND EVEN AS IT HATH TAUGHT YOU, YE SHALL ABIDE IN HIM." Please see how it is *impossible* for HOLYSPIRIT to teach souls differing interpretations of Scripture. By all righteous understanding, such duplicitous belief, equals division; it is two faced and not of GOD. There are endless false doctrines taught by souls falsely claiming the indwelling HOLYSPIRIT; yet only *one* shared firm truth, courses through those truly filled. If a doctrine is accepted by worldly souls, it is a worldly doctrine. Modern Church pulpits are filled with self-taught false-prophets, and seminary graduates, taught by men how to financially fill a church with compromised worldly souls, having desires for an eternal life of comfort and beauty that please the flesh. These lukewarm souls, live by false beliefs, traditions, pride, and envy. Denominational leaders, bow to error, weakness, and sin, as much or more than their students. Matt. 15:12-14, "*THEN CAME HIS DISCIPLES, AND SAID UNTO HIM, KNOWEST THOU THAT THE PHARISEES WERE OFFENDED, AFTER THEY HEARD THIS SAYING? BUT HE ANSWERED AND SAID, EVERY PLANT, WHICH MY HEAVENLY FATHER HATH NOT PLANTED, SHALL BE ROOTED UP. LET THEM ALONE: THEY BE BLIND LEADERS OF THE BLIND. AND IF THE BLIND LEAD THE BLIND, BOTH SHALL FALL INTO THE DITCH.*" JESUS says here that many religious souls do not bow to or hold FATHER'S beliefs. Leaders *and* followers who desire and choose false doctrines, will be denied entry to GOD'S Kingdom. CHRIST urges us to focus on seeking and living the truth, and pass by all who argue to alter the truth of Scripture. I have visited many churches, and heard weakness and error to various degrees in each one to date; this should not be. It is the true beliefs *within* the walls that make a true place of worship; and not the walls and people themselves. We know in the beginning, FATHER went to great lengths to create *his* ways of lively spiritual and physical perfection, in the hope of providing an eternal home for many souls finding eternal love and peace. And FATHER'S great hope is to love, and be loved eternally. So, what is our core love, and more importantly, what can our love truly become? All gifted souls within this fallen physical reality are capable of a lesser love. Yet most souls appear *in*capable of wanting to nurture, grow, and protect love until it rises to its eternal Godly form. They can't give their greatest measure to GOD, and don't seem to have eternal love for anything. So, the tiny warm ember of love that most souls are capable of, is a lesser fading hope of what love is meant to be. Each warm ember of love, was meant to become an eternal white-hot flame. Eternal love is not defined by any individual; only GOD defines it. Reprobate individuals can love people, things, and ways, for certain durations; but never eternally. Fewer and fewer worldly loves endure for even a year, as more and more end in disillusion, indifference, and hate. How can *eternal* love, turn to hate? We see true marriage decreasing, as divorce and vile false marriages increase. Children are indoctrinated against GOD'S ways, leaving families unable to truly communicate, or love. Liberals cannot defend their immoral positions, as GOD'S created nature of life is to be fruitful and multiply; remove that, and all life will die out. Perverse souls do not perpetuate life; they abuse and destroy life for their ungodly pleasures. The perverted can never in honesty, debate GOD'S words; they can only attack them. Divisive liberalism is pushed by liberals into every aspect of life. And by the natural laws of progression, imperfection begets more imperfection, and sin begets more sin. Did not that one first imperfect desire of Satan, grow to fill creation. Did Eve's first sin not quickly spread to Adam through his love towards her. This world is over-flowing with sin and sinners, being accepted and embraced by churches without repentance. And at great cost, sin consumes all it can. Only perfect things are eternal; so, if a soul says no one can become perfect, they deny the power of GOD'S eternal love. They are currently of the spirit of antichrist whether they know and admit it, or not. Eternal love must be founded in a desire for FATHER'S perfection and ascending order of souls under GOD'S eternal ways. The lawful attributes of created probability and chance, seen in all true free-will choices among many, say not all souls will love GOD'S ways. This immutable law of free-will souls, is loosed in a wondrous shared existence of true choice, to allow and see what each soul will love. And in that GOD is real and loving, GOD abides by this law. The Scriptures declare that after the fall of mankind, FATHER saw in grief, that most of mankind are never going to desire his eternal ways; and in these last days, we see this is true. So, we must know and remember why and where this eternal divide between the lost and the saved is placed. *Is* it placed by GOD, or by free-will choice, or both? Is it many, or a few that cross that divide between temporary and eternal love? There exists no way for love to be imperfect and remain eternal. The term "perfect" in its association to eternal love, means a soul's pure love born through faithfully living in GOD'S omniscient ways. In this, GOD has promised all who come, an eternal rising joyful loving peace. Why is this so hard to choose for the many? Because their great core desires are for the divisive pleasures of this world. So, these teach, urge, tempt, and shame each other, into not believing or living in GOD'S perfect ways. Most *professed* Christians, think the only way to have a devil inside of you, is to be possessed by a fallen angel. But if you deny living in the created ways of GOD, and live by your own beliefs as the devils do. GOD says you *are* a devil yourself. Jesus said to Peter in Matt. 16:23, "*GET THEE BEHIND ME, SATAN: THOU ART AN OFFENSE UNTO ME: FOR THOU SAVOURIST NOT THE THINGS OF* ***GOD***, *BUT THOSE THAT BE OF MEN.*" In a moment, Peter stopped being a lesser spirit son of GOD, and became the contrary spirit of a devil. Yes, he repented and was restored. Most by their core desires must be the god of their own ways, they can't see that GOD'S created ways of life are perfect, and theirs are not. The number of reasons why *individuals* love and choose this world and self above GOD, are then equal to the number of paths leading to the abyss of hell and death. A great part of the essence of eternal love, is in our choice to never hurt, deny, or unnecessarily burden those we eternally love. And we do this *for* GOD more than all others, not *less*. GOD'S work to sustain all life and all things, is laborious. And our love is seen in not increasing that work beyond GOD'S need and desire to do so. If a soul makes it hard, expensive, and hurtful for you to love them, you know exactly what that means. It means you are weakly loved, or unloved and abused. By the truth of GOD being the only truth, we see GOD'S conscious formula for the only good real and true strong-emergent that exists; being something that never previously existed, coming to exist by no physical cause and effect, is a pure sacrificial loving desire for eternal life. Forceful desire is the only true emergent in reality; the only thing or concept in this created physical reality that can ever *newly* exist. The only thing that comes in its fullness, from nothing else, is our desire. So, we see new and old, and layered preconditioned desires, we see spiritual and physical desires, we see desires of true need and false want. So, if souls desire to live contrary to the created ways of reality itself, how can they live the perfect ways of eternal life. Consider again, that if all there were to choose from, were only the good created sanctioned things of GOD, so no one could choose anything other than GOD'S ways. Would there be any true choice, could love grow to be real? No. Must FATHER say, do *not* choose at least *one* of two desirous ways of life, for our free-will choice of love towards GOD, to grow and exist? Yes. And we see that GOD denied just one choice among the abundant perfect choices in his Garden; one distinct denied perfect tree belonging to GOD alone, being denied to Adam and Eve. So, their taking of it would be a choice of theft, and a betrayal of eternal faithful responsible love. For free-will love to truly exist, the free-will betrayal of that love must be possible. Must this same choice and temptation be in the New Heaven, *after* already proving our great love? No, not anymore. GOD'S family of eternal love will be found, proven, sealed, and eternally cherished. A choice for evil will not exist there anymore. GOD will never again create anything we cannot have. We see this choice of love put before Adam, and later to Eve, came with GOD'S dire verbal warnings, that this one choice must be shunned. And it being shunned, was evidence of Adam and Eve's perfect desire to be and stay one with GOD in their created ways and provision. One choice to not hurt, deny, or betray GOD, had to be set in the midst of that abundant allowed perfection to find love. Love and hate now flow between mankind in our many choices that affect each other. Adam and Eve had a choice between something they could not have, and GOD; a choice to stay, or deny and leave GOD'S loving ways. And that *first* imperfect choice, opened an endless hell of every imaginable evil choice to be desired. And every evil choice, makes FATHER'S creation more imperfect, requiring much work to make it perfect again. But FATHER does not increase his original level of sustaining work; so, his care grows more insufficient as imperfection is added. By all introduced increasing disorder, the eventual loss of the physical body awaited Adam, Eve, and all who follow. The separations of hell and death had entered into all mankind's conscious and physical equation. They had never projected a forceful rebellious formulation before, or knew of all endless hell and death. Did Adam and Eve immediately feel sick? They surely sensed their life was spiritually and physically fallen and different. They felt loss, weakness, and new pain, as they had fallen from a high perfect light, to something far less and imperfect. In all their omniscient learning from GOD about their gifted soul, and created body, they were told that one evil choice leads to death; and death is the loss of their perfect spiritual and sustaining physical connection to GOD. Our soul is ours eternal; but a body, being the *required* half of our shared Dualist existence, is a conditioned provision. And in having chosen to open and enter the door to hell and death, mankind was put out of the perfect garden. So, are Adam and Eve eligible to repent and be saved; seeing they were created in perfection, and walked and talked with GOD in person, as the fallen angels also did? Adam and Eve, unlike all mankind who follow, were gifted, created, and set upon the pinnacle of true choice, and not born on the wrong fallen slow small side of choice as we all were. So to me, their eligibility for grace is an unanswered question. And now, many don't love GOD at all, and many more hold imperfect insufficient loves for GOD and their ways of life. Yet there are the fewer who do find and grow and hold such a faithful sacrificial love, allowing them to hear the truth that brings a firm eternal quality to life. These fewer *do* truly love the eternal ways of life. Adam and Eve had to protect their love for GOD from *one* evil choice. We see mankind's individuals, vainly try to protect their weak temporary loves from the relentless things that attack those dying love. No soul can grow or hold an eternal love through the fallen imperfect evil ways and beliefs of this world. Our desire to live with GOD eternal, rests entirely in coming out of all the fallen physical ways of this world, and returning to GOD'S perfect spiritual ways. If we love anyone or anything as much or more than GOD, our eternal love is broken. If we want to eternally hold onto all that is good, we must hold onto the one belief and understanding of GOD, and love the things of GOD; or we must drift apart forever. Everything, including our soul, is a gift, a sacrifice, and a labor from GOD. And to love any gift above the giver equals the confusion of hell. Every choice falls there in front of us, and the evidence of what we love most is seen in every choice we make. Who do we love? Do we hold GOD'S eternal spiritual well-being above all others, and all worldly needs and desires? Our familial love must be of one mind and accord under and for GOD; in this, we add to the Kingdom. In each souls created expanding moment of existing, we all choose most, what we love most. You cannot deny that you love most, what you choose most. What do you love most each day? If it is your earthly family, this is wrong. Such individual truths and loves, are not eternal truths and loves. We see endlessly changing individual truths, and we see GOD'S one never-changing truth. We either want GOD'S belief and eternal ways of perfect provision, or we desire individual unsustainable ways and belief. Every evil desire, is a wanton disobedience towards GOD. Consider a created inherent weakness in mankind, being we cannot create or sustain the physical half of our existence. So, separating one's self from the sustainer of reality's physical half, brings only hell and death. Again, hell is an ever-inflating abyss of confusion that divides souls from communicating with GOD and all others forever. And death is our conscious/physical separation from GOD'S *good* provision. Our first death, brings the loss of this body; *death means separation*, full *stop*! All who walk the earth in their own chosen confused beliefs, are already unknowingly walking around in their chosen separation of hell and death. JESUS said in Matt. 8:21-22, "*LET THE DEAD BURY THEIR DEAD*". Many have, and more will be resurrected back to GOD from their fallen spiritual states of death and hell. But *all* souls will be physically resurrected from the first state of physical death; some to eternal reward, some to eternal separation. GOD warns, that in choosing individualism, we are cut loose from the mooring truth of GOD, to fall into a dark conscious pit, minus all pure eternal understanding. Most don't grasp the need for our *eternal* mooring. They don't see that letting go of GOD'S truth belief and understanding, at first feels like consciously floating free, and flying a little higher and higher for a number of years. But then one day, you realize you are slowly and eventually drifting out into the dark loneliness of conscious space, where everyone consciously drifts farther from all others in belief and understanding. And with no way back to GOD'S mooring one belief, this brings an eternal confused loneliness that can't be imagined. All *eternal* souls require a shared *eternal* truth to hold us together in eternal love and communication. This cannot be logically argued against, so please don't try. In making continuous godless choices, each soul increases their spiritual hell and physical loss. Most do not sense they are separated from GOD, or what that means. And when they do find out, and it's too late, they will wail and nash their teeth in their fearful anguish. Individuals live for temporary distraction and satisfaction through physical and dark spiritual desires. But these evil acts do lose their pleasure and grow old, as all temporary excitements turns to tedium. Perfection has its *eternal* joyous outcome, but imperfection has a continual suffering demise that can never end. Do not be fooled, there is no middle ground of being considered unaccountable once past the age of naiveté. From our first knowing of good and evil, we are choosing *our* way, or GOD'S way. Whether you see and believe this or not, is your gifted created purpose. A soul's false consciously formulated structure, can grow quite complicated in layers of built-up denial and supposition; and each individual remains a unique equation. It is literally a god complex, to want and need to form one's own belief. Confused souls think they take something good, denied them by GOD; but it is their choice for evil, and that choice was first given to them *by* GOD. Many in pride and envy, see choosing GOD as a coerced choice. For souls with such beliefs, the only way they could choose GOD, is by being forced. But GOD can't and never would, force anyone; so, sadly, GOD must let them go. It is an eternal law of free-will choice. All who choose GOD are never forced. GOD created the means and possibility to allow every choice, but only we desire and make *our* choice. Some, in a fairly strong yet insufficient way, love GOD; and these live in ongoing efforts of failure and shame as they never seem to overcome. In hearing this, do not *under* or *over*-estimate the great mercy of GOD. If a soul holds to any enduring love, and fights, and has not committed the one unforgivable sin, GOD still helps, and hopes they overcome their wrongful desires. Individuals rely on this world's false knowledge, so how they see reality is false; yet they can still see their self as wise and intelligent. If a soul can't spiritually *and* physically describe an *eternal* loving peaceful reality for many, they hold no true science or spirituality. The world is pressurized more and more towards implosion, by forces that cause misunderstanding and separations, as all self-righteous anger grows between individuals. As insane division grows within all groups, souls become imprisoned more and more as one unto themselves in their own slow small individual understanding and perspective. This is the foretold never-ending hell of reprobate souls, unforeseen by mankind; yet foreseen by us. It's not what we each physically see and hear, but how we spiritually and scientifically interpret all things to form *our* soul's believed reality. Most all now perceive everything differently to a degree. Even if many agree on something broadly in general, deep down, their purpose and desire for all things in some ways differ, change, and conflict. All souls come to believe or disbelieve what they see and hear, *only* after it is filtered and interpreted through their current desires and beliefs, to become what it means to them now. We each see the world in the exact way we desire to see it. Once again; as we are, is how we see it. If GOD'S ways, truth, and belief, are ours, we see *all* things through GOD'S perspective. And if not, or if one's soul is partial in this, one will interpret GOD and this world to allow degrees of individual ways, beliefs, and understanding. They will put their spin on it, and it will be false. The original authored eternal way that things were created to be by GOD, has been lost to most all of mankind. Mankind's fallen ways sustain nothing; so, their ways will never be allowed into GOD'S New Kingdom. Rebellion, means defiance and resistance; can you correct your internal rebellion and its resulting destruction? Yes, if you find the strength of eternal love. If you are turned to believe *a lie*, is this rebellion? As soon as Eve and then Adam rebelled, it was by their new individual false belief. All individual rebellion, is born deep in the desires and belief of the beholder; and FATHER'S perspective sees all rebellion. So, to see all your own rebellion against GOD, you must see the fullness of GOD'S truth. If we only see half of GOD'S truth, we can only see half of our rebellion; and only half of any new oppositions taking root. For any soul to follow another to any *degree*, requires an equal *degree* of agreement in desire and belief. To what degree do we each follow GOD? GOD says we are all learning vessels of desire and choice who accept or deny all things that come before us. So, it is wise to flee all temptation and evil influence, to not stay among them, and to run to GOD. To turn from all rebellion, we must remove every evil formulation of belief from our soul. Again, among the lost, there are what seem to be strong surface agreements and alliances here and there, but they are superficial, vague, and weak. We have all held evil attributes to differing degrees; but the fewer strive to replace them *all* with the attributes of GOD. And in the end, it is not us, but GOD who sees and judges the difference between the two. How easy it is for most to not see, or forget we are forceful Godlike creative souls, who effortlessly desire and create our streaming multi-sensory particle-wave-based perceptions of brightness and life. We all live in our own measures of hope; and among all individuals we see a physical compassion. But they have no faithful responsible spiritual compassion for every soul's eternal spiritual health. Around the world, we see all spiritual good desires growing colder in the church; and when no new eternal love can be found by GOD, the end will come. Rev. 20:11. *"AND I SAW A GREAT WHITE THRONE, AND HIM THAT SAT ON IT, FROM WHOSE FACE THE EARTH AND THE HEAVEN FLED AWAY; AND THERE WAS FOUND NO PLACE FOR THEM."* We must see in all things, there are the righteous sustainable ways of GOD. We see one responsible faithful Holy love; and endless irresponsible, unfaithful, unholy, loves. If you believe GOD should have physical compassion upon sinners who desire to be trapped in endless destructive ways, against the ways of GOD, and allow them into GOD'S Family and Kingdom. I see only the destruction of GOD'S Family and Kingdom in that. I have no desire for such false compassion. To be eternally kind to any evil that brings hurt upon all, is false compassion. The hurt of evil is then spread on all; how is that compassion? Where is the compassion for GOD'S Children? Compassion is never for the lost alone, or there would be no final eternal divide. I will in a cautious way, feed, clothe, and supply a lost non-reprobate soul, three days journey, and share GOD'S words of hope; and hope they return as a Godly soul. I send no one else to do this. I see no righteous compassion in putting any child of GOD in danger. I send no one to help another soul who may see them as prey. My compassion for GOD'S children, outweighs compassion for the lost. I hope the wicked soul will change; but not at the loss of any Godly soul. Will FATHER compassionately allow Satan into the New Heaven? GOD forbid. We are called to righteously go two by two, but only in GOD'S mature strength to do only as GOD commands. A reprobate will manipulate each word's meaning to prey on every weak soul's tender emotion; they turn souls striving to get out, back into the world. Individuals live at a threshold of losing nearly all ability and hope to ever see or hold the truth of GOD. I can watch hours of televised news, and sermons, and never hear one *pure* statement. I hear a few truths mixed in among their confusion. My statement of hours of constant misinterpretations and lies, accurately describes the current unrighteous propagation in this world. A weak truth of GOD does not exist; there is the one pure truth, or it is a mixed lie. There is no hope of enduring peace or love for mankind, led of themselves. There are a few strong souls who might slow or bring moments of rest and hope. But upon their effort, the world will surround and use every lie to destroy them. Most soul's description of peace and love for all mankind, align to their own desires. One soul's call to peace and love in this world, equals another's call to war. One person's greatest desire, is another's greatest aversion. One person's greatest love, is another's great hate. So, how shall these find peace together. Can they all compromise to find a common state of misery, in losing who they are? How can souls compromise their greatest love, with their greatest hate? FATHER'S truth and ways are designed for eternal peace and love. These can never come from mankind; they can come only from the author of reality. Do you see why FATHER repented making mankind. FATHER lost one third of his Angels, and it is not fully determined what percentage of mankind will be lost, but it appears to be above 99.8%. Our spiritual study and accurate understanding of why one third of the angels and most all of mankind, fall to be individuals, is so important. It is by all these things that the separations in conscious spiritual existence, and physical reality, are occurring. So, we see that what, when, how, and why we learn, is crucial. Be taught and teach your children well about GOD from the moment they are conceived. Fill them with understanding of GOD, and they will learn it. Do not irresponsibly say it will be their choice when grown. Satan does not wait, and by then Satan will firmly have them as perverse zombie reprobates. How has this individualism entered so deeply and broadly, creating all these weak denominations and sects in GOD'S truth? It is by those who declare with their *mouths* how they love and follow GOD; yet they love this world and the flesh more. Evil enters all acts of life by the desires of individuals, demanding to be accepted by all, in what they choose to be and believe. We call this "The Law of Propagated Desire", as it compromises weak souls. We see in weak believers, their strange innate desire to accept the confused hope of all souls being possibly saved. Each sentient soul, is a conduit of chosen understanding, truth, and belief. Souls are other-worldly spherical vessels who come in contact with true and false conscious formulas of desired understanding and love. And individuals interpret all true and false understandings through a personal biased conscious formula. GOD declares who are good and evil souls, by seeing their chosen desires, perspectives, and beliefs. Confused evil souls believe no one can be perfect; so, in an inadequate way, they think they are good enough and acceptable. Self-preservation, being their inherent conscious formula, blinds most souls to their own divisive nature and future. Though *I* strive to hear, see, and live, in a singular desire for GOD'S perfect ways, *I* too dwell in my hope and faith to be found on the side of GOD. We live watching and hearing a fallen world of evil desires and pursuits surround and strive to pull us all down, to make us all as they are. Lucifer, and around a hundred million fallen angels, and billions of fallen mankind, are very real; and they blindly seek out every weakness as a point of entry. Satan and his legions lay subtle traps before the strong. They strive to desensitize and lure us all, inch by inch, towards our precipice of sinning and falling. Souls are eternally defeated through a multitude of tiny losses. Evil can rarely lure a soul across a great distance in an instant, as most find it greatly discomfiting. But slowly and relentlessly using the world and all within it, so subtle, patient, and prolific, are evils methods. You look this way in this world, and there is evil; you look that way, and GOD'S ways are not there. We hear from one soul, and then the next, only twice the delusion. Scripture says Satan comes disguised as an Angel of Light, declaring and showing himself as God; how sinister. Satan does not work in pubs, brothels, and other dens of iniquities; these souls already work *for* evil. Satan's work is in the pulpits, schools, and governments, at every *divide* of weak and strong spirituality. Once souls are separated from GOD, they continually, yet unknowingly pull each other down into hell and death; this then becomes their unrealized purpose to exist. Godly efforts to correct a soul's beliefs and doctrine, come under attack by other souls wanting their liberal desires and ways to be allowed and included in GOD'S ways. Most souls in pulpits, succumb to evil pressure; but I assure you, GOD does not succumb. Satan sees and uses each soul's own wrongful desires against them. If our great desire is to be led by GOD to find and help others, *all* the way back to GOD, and not partially back, this is our only good work in this world. GOD'S work requires being properly strengthened, taught, armored, and led step by step, *by* GOD. The whole world is an evil trap, set to use our every weakness against us. Even our zeal to do GOD'S work can be used by Satan. We can never fixate on bringing certain souls to GOD, who says speak two or three times to the lost, plant a seed, and move on. Another may speak to them in a moment they might hear; but if you obsess upon being heard by certain soul, Satan will use them to pull you down a little at a time into their lost reality. Never over-estimate your strength, and guard your soul, or you can lose ground. Never expect any reward for Godly work, as this lively work *is* itself, is our reward and purpose. Live this way, being happy in doing all we are created to do, and being all we were created to be. In attacking us, Satan may use everything; but in being truly in GOD, we are at peace. Evil, not having GOD'S or our permission, can do nothing. Being GOD'S true child, Satan cannot lightly touch the body or tempt our soul, minus permission. Now if we love a lost soul too much, and can't accept their eternal loss by their free-will desire and choices, we are on Satan's hook. Satan can lure us, through our resistance to accept the realities of free-will choice. Never ask GOD to save a soul against their will, or change their desire. GOD can love, hope, call, and touch. But beyond that, the step of turning to repent, is solely in each soul's hands. Many souls are lost, and many say how could GOD allow this? But how could GOD stop it? Do you see around 4,000 years ago, every soul but 8, on the whole face of the earth, were reprobate. Scripture says FATHER was grieved he created mankind, and cleansed the earth of all whose minds were on evil continually; all but 8 souls. GOD cannot over-ride the choices for what a soul loves most in their own existence. Accept this absolute created free-will reality here and now; or hold a fatal flaw of belief that Satan will use. FATHER hopes *all* souls would desire and choose to be one with GOD; but in all evident wisdom, GOD has found this will not be. In gifting multitudes of *free-will* souls, FATHER knew that probabilities and uncertainties in true choice, were set beyond his control or knowing. And with each new soul and new sin, the probability and uncertainty of loss increase. Each soul's final freely chosen state, is not part of GOD'S omniscience. Each free-willed soul can desire and choose to self-rule and self-define what they see as good and evil. Our creative ability and our sensations of physical pleasure are a weakness for mankind; but in ignorance and weakness, why would any think we are wise, and refuse GOD'S truth? Each soul's awareness of understanding, belief, and love, just is and will be, what it freely is and will be. We all desire and believe what we freely desire to believe. We are gifted to choose who or what we love; for only in this way is *our* love our own creation. Why does FATHER not stop right *now*, and separate all who do evil, and put them in their eternal prison. That, he will soon do, as the lost self-righteously howl, why does FATHER cast us out. Those who say GOD created me this way, are insane; they created themselves that way. When a soul sees themselves as the center of all reality; in looking to the right, they see others are too much, and looking left, they see others are not enough. These souls compare everyone to themselves and everyone else; where all who have the eternal perspective, compare ourselves and everyone to GOD. It is GOD alone who says who is eternally loving, or not so. And in all their ever-changing individual hells, love is a painful illusion in a mirror. Individuality *is* this world's most prized possession; but at their last, they will see how their ever-increasing individuality becomes their own created eternal isolation. They think individuality, takes them to great heights; but it is an anchor around their neck, pulling them down into hell and death. It's a growing void in their understanding of their own chosen darkness, consuming their light. It is their chosen eternal anguish and sorrow. The old saying, the heart wants what the heart wants, is insufficient and inadequate here. A sufficient adequate utterance is, a heart that truly *does not* want GOD, *cannot* want GOD. A soul who throughout their hellish life here, never desires the fullness of GOD'S love, cannot at their last moment, suddenly desire the fulness and love GOD. At their last, in fearful self-preservation, many can in tears *say* they now love GOD; but how can this be a true eternal love. Know, understand, and see, how eternal love now grows. If a soul asks GOD to wait till later, GOD is not loved. It is GOD'S design and immutable promise, yet most do not care that love is meant to be eternally cherished. Most on balance, take love for granted as an aspect beyond their control in all their worldly lives. Most souls on balance, use, abuse, and expect love to some extent, for self-satisfaction. Falling short of eternal love, brings only pain and loss. Love is such a wondrous marveled force. It is only our desires that align our two halves and bring our awareness of existence in GOD'S reality. There is joy, peace, and enduring reason and faithful purpose, that the oneness of eternal love alone brings. And great anguish, loneliness, and separation comes to all who do not find, or lose their eternal love. Love is not something that just is, or is not; love resides in strength or weakness, in purity or impurity. And if unguarded and unnurtured, love will *always* weaken and die. We see Heaven filled with souls who treasure, guard, and nurture GOD'S and each-others love eternally. In Heaven, all souls will go beyond the protective gate and walls of New Jerusalem to mourn the loss and cost. There, seeing all anguished embodied souls roiling in their confused abyss in the Lake of GOD'S Fire. Why do most souls who fill the churches in these last days, believe FATHER designs an *individual* plan and purpose for each soul, and not one perfect uniting plan and purpose for all his Children to find by becoming one in him? Because they cannot let go of their individual desires, beliefs, and purpose, forming their religious individualism. These think their own chosen individual life, is GOD'S individual predestined plan for them. The fallen Angels, and the fallen of mankind, have denied and altered GOD'S ways beyond recognition, and resist returning to those ways. Most all mankind say they want eternal unity and love between all; but in their hearts, they want that unity to align to *their* individual desires, and fulfill their every want. This fact shows the impossibility of Heaven existing among religious individuals with differing desires and beliefs. If you want some of GOD'S ways mixed with some worldly ways, can the New Heaven be a perfect place for you? Or will it lack certain things you need to be happy? Thinking of heaven as a purely physical place, that GOD will make perfect for each of us, is a fleshly belief. GOD is the author of one sacrificial conscious loving peace, and one good eternal spiritual and physical existence. It is all our errors of individual desire that must change, and we must desire and conform to the ways and truth of GOD. Heaven is greatly spiritual, and minutely physical; and most believers know they are not one with GOD, yet think they are honest and religious *enough* for GOD. These believe their ongoing sins will forever be forgiven in heaven, and that if their good efforts in life out-weigh their bad, they have met GOD'S standard. These fail to grasp that any and all eternal good, dwells in eternal perfection. All that is thought to be good by mankind, outside of GOD'S omniscient authored ways, is corrupted with evil. JESUS said there is none good but GOD. This is the true and eternal formula for love and life. The only way to be or do good, is to *emulate* and live within the ways of GOD. It's such a strange thing, watching many in their confused differing efforts, trying to do what *they* see as good, not seeing our souls are insufficient and incapable of knowing GOD'S good on our own. Yet we are capable of being taught all good things by GOD, *if* we so desire. Can we try to keep certain fallen souls within the reach of GOD? We can try, but the outcome is doubtful in a world ruled by individualism. We see individualism brings struggle to all concepts and needs of love, life, and governing. A uniting Godly government is needed, not an ungodly government. The few more moral Governments found in history, were a few ruling Priesthoods, and a few Righteous Monarchies. Of the rest, the closest any came, was the original, once more free and more moral Representative Constitutional Republic of the United States under GOD. But by the U.S. enshrining more and more rights to ungodly individual perverted ways, the U.S. devolves towards an immoral *godless* Socialism. At our founding, our constitutional obligations were incomplete among its citizens, and good corrections were made. But an ungodly immoral sense of freedom has seized our Nation. The U.S. Constitution's Federal powers were very limited for good reason, but by a changing powerful world, those limits have been altered. The far greater power was given to the States and their representatives of the Republic to better preserve our moral freedom; but this too, has been altered. Our Constitution, requires Godly wisdom, courage, and a knowledgeable responsible citizenry; and this too has been altered. Our Constitution requires a true historical education for all, built on Godly morals and our enumerated principles; this foundation has been undermined. Our religious freedom enshrined in our Constitution, is required for a Godly moral religious responsible people to live under the authority of GOD; this freedom is being diminished. Our Constitution's Religious Exercise Clause, is very clear. So, for any federal, state, or local Jurist to decide *any* case, centered on Free Religious Exercise. They must know first and foremost this is a question of what is religion; and secondarily a question of what is freedom and lawful exercise. This free exercise clause, applies to *historical* long-held long-shared beliefs, orders, and traditions that souls know must be propagated and passed down. A Jurist must intimately know what the term religion entails and extends, and I say they must be a firm believer. The word religion in its true definition, means those things that historically (throughout millennia), souls for great reason and cause gather together to believe, declare, and vow they must receive, give, hold, and live. Being a true ancient definition of good shared lives, lived to their ultimate purpose, in honoring it all, under the higher knowing source of it all. The modern religious concepts of materialism and humanism, are vague religions; but they are not historical, or constitutionally protected. New cult or followings cannot claim Constitutional protection. New religions and Court challenges are designed to undermine Historical Religion. True Historic Religion is constitutionally protected in all public areas and places of residence or worship. Two souls holding differing historical faiths in all their ancient ways, have all public rights of speech and propagation in the public square, that stop at open intimidation, obstruction, menacing physical harm, coercion, or disturbing the peace. Freedom of exercise, means the unhindered non-violent non-coercive exchange of ideas. Religious beliefs cannot be privately, publicly, or physically imposed upon another; and they cannot be privately, publicly, or physically restricted by another's beliefs. Now yes, the public square includes public schools, which cannot be claimed by atheist as their bastions of indoctrination. Freedoms enshrined as a Constitutional Representative *Republic*, are built upon a truth that to allow differing beliefs, you can't govern belief from a central form of government. But at the State levels, they can be far more discriminate towards belief, giving its religious majority, more control. And at the Local levels, the broader religious majority view is yielded to a little more. Now this always, as everywhere in the world, has a minority belief that is not served in the same way as a majority. Yet their freedom of belief, practice, speech, and moving to a place more in line with their beliefs is allowed and encouraged. Communism or Socialism are built upon the idea that it doesn't matter what the local or over-all majority or minority want. Governance is controlled by an Elite Bureaucracy of like-minded godless individuals, who rule strictly by statistics. Such rulers have a god-complex and godless-mindset, as they demoralize, indoctrinate, and force one belief of a perceived good, upon all citizens. By this, most every soul is separated from GOD and lost. The spectrum of Democracy, Socialism, Fascism, and Communism, form the one-way hallway, that leaves the room of moral religious freedom, and leads to the room of total demonic global controls (the Beast). The world has fallen into sinful disbelief, deceit, immorality, propaganda, and hell. We are seeing the Marxist ideology, infiltrating all aspects of American life. And Mainstream Marxist Media (MMM), portrays all true Godly belief as racist, bigoted, and extremist. Every disparity sensed in every minority is now used by Satan, to force the majority into subjection. These D-zombies are incapable of recognizing good and see only a narrow-indoctrinated understanding of injustice. They hold to a highly amplified understanding of a narrowly focused good they might do, as they become consumed with bringing an end to one sensed injustice. These feel compassionately superior to those who do not approach singular injustice in the over-amplified way they do. When one is infected or possessed by a zombie (singular) desire, it presents as a viral cognitive passion. A cognitive zombie, is a soul so singular in one cause, desire, and understanding, that they want to devour or destroy the mindful process of anyone who does not think like they do. You can rarely reach or reason with a social D-zombie through debate or dialogue, as they are programmed to shut down debate and dialogue. They cannot see how in not considering *all* injustices equally, that their remedies to ease one injustice, cause greater injustices to all others. They cannot see that focusing on one aspect, allows all other problems to grow worse. If a house needs over-all maintenance, and you foolishly focus *only* on changing the air filters several times every day to have the cleanest air possible. Sooner or later your house falls around you. I was taught by all understanding, that a wound never heals if you obsess and pick at it. All change requires its natural moment of forgiving, forgetting, and rebalancing. Nothing existential can be fixed faster than it can naturally be fixed. And nothing is truly forgiven until it is spiritually forgotten. A job rushed by wild ill-equipped workers, is never done right. A narrow and frantic mind, is incapable of seeing the broad ways and means required to fix everything together as one. All rational complex loving souls, see the way to ease all societal ills together. Cognitive D-zombies believe imperfections must be fixed one a time. These are *illogical* emotional extremist, driven by irrational beliefs, fears, false compassions, and an evil immoral indoctrination. D-zombies are taught to hate every command in Scripture; and for the most part, don't realize they are Satan's Children. Now by Government, we see the forced collection of taxes and their allocation to causes that many know do not strengthen family or country, or preserve the liberty and dignity of all its citizens. The freedom loving right, see the needs in their lives, yet see their resources taken and given to evil causes, as the needs of their children are denied. In this these D-zombies subvert the truth, and their desire for their perceived justice, increases all forceful divisions. The left divides the country by amplifying and focusing on perceived victimhood and oppression. This is the bait on their hook, as they fish for overly-emotional, irrational, illogical followers, whom they then drive head-long into insanity. When two souls disagree, they each think they are correct, and the other wrong. They never consider that most always they are both spiritually and scientifically wrong; and agreeing in error, makes nothing right either. We must yield to GOD'S perfect ways. What is the conscious formula for modern politics. There are endless examples where mankind's politics try to make things better for some, that never bring good results. Liberals blindly group and label people, claiming others who see a holistic approach are evil. The D-zombies does not understand, see, or *care* about the injustice caused to others by their divisive ways; they just divide people, and push the world towards its coming end. They don't care that in taking a burden or debt off of one back, that burden or debt must be set down upon another. This is the nature and definition of a burden or a debt; it is to be born and paid. Even our debt of sin cannot just disappear, except by CHRIST, who paid dearly! Who do mankind think they are, where has a true good been achieved in governing as if they are God. If you have ever thought you saw a true measure of justice in this world, delivered by mankind, I promise the resulting unseen equal or greater injustice is there, passing painfully through the life of others. The only pure physical good that can be done by anyone, is when one soul led of GOD, *personally* by loving choice, bears a burden or debt of another, being our second lesser part of love. But taking a burden off one to unrighteously place it on another, is a self-righteous act of arrogant godlike pride that breeds division. The harder the world tries in their endless confused ways to make it better, the more divided the world gets. Yet they can't see it happening, it is all too big and too fast for them to see at once. The only physical wisdom mankind can find in this world, is that this world can only offer the sum fullness of all that is created within it. And as it turns out, by the imaginations of creative souls, longing for physical comfort and longevity, we see there has never been enough wealth or production in this world to give to everyone, the standard of physical life that this world can offer the few who have attained so much. Name one form of governance that has reduced the injustice of one soul, without increasing the injustice of another; you cannot, ever! No matter the form of government we see in this world, no one can remove old disparities, without creating new disparities. The greatest created disparities always come from centralized intrusive governments like socialism, communism, fascism, and pure democracies. We see all dictators around the world become billionaires while in office, as their masses dwell in shared misery and hopelessness. We see in the west, that greed, immorality, and insanity, have increased in a huge number of politicians. These seek and learn ways to gain wealth and hedonistic pleasures through manipulation and influence. Politicians stand up and cry that lawful Corporations and their Chief Officers, reap too much reward, as these CEO's create monetary increase for *public*-investors. Greed does exist, and always will in this world. Yet the greater greed and corruption exists in governments who control all things.

Power always corrupts; and absolute power, always corrupts absolutely. You will find far more corruption in centralized government controls, than in any free-market, representative governance. The U.S. Government literally *loses* track of billions of tax dollars all the time through programs, and no one is held accountable. But those running publicly owned corporations are always held accountable for unaccounted losses. But governmental control is accountable to no one. In the shadows, politicians gather wealth without any lawful labor or enterprise, yet point their fingers at those who gain through true labor. Politician's monetary gains are acquired through non-public knowledge and trading influence toward fashioning regulation and law, and by politically dividing people into groups against each other. Most souls do not see Satan and his Children going about their task of dividing all, against all, using all temporary illusions of physical comfort, pleasure, and safety. Where do we find the greater numbers of those possessed or influenced by demon spirits. There are the lower rag tag demon spirits if you will, selfishly praying on weak souls, seeking to sense perverse physical pleasures. But Satan's hierarchy of confederates seek to control and or influencing Globalist Politicians, and the Worldly Religious

Hierarchy. Are all Globalist and worldly religious leaders, demon possessed or influenced? We can't know that. But I eye all Globalist, and Religious Hierarchy with a discerning eye, as they contradict Scripture and fight the fulfillment of GOD'S Prophecy for his Kingdom and Israel. These liberal elites, whether truly demon possessed, or just influenced, all Globalists fight against GOD'S Kingdom. Globalists are humanistic, hedonistic, communistic, and materialistic at heart. Globalist will deny having most of these attributes, and all will deny having at least one. Yet if you study and compare not their words, but their actions, you will see all globalist exhibit these traits. The conscious formulations of belief that enable their ways and actions, do reveal their common antichrist ideals. Before any get riled at this antichrist accusation, let me explain. CHRIST said you are for him, his ways, and commands, or against him. If you are not living by his true and complete teachings, you are by *his* definition, acting anti, against, and contrary, to Christ in your certain ways. Yes, there will be the embodiment of one fallen angel called the Antichrist, and there are degrees to being of an antichrist spirit, but there is no grey area of obfuscation. All globalist's share the goal for all souls to be under one strong godless central world elite rule; called the Beast in GOD'S Prophecy. This Beast will control and coerce all lost souls to worship power and physical pleasure, and see Satan as GOD. These seek to steer and organize the selfish desires and self-righteousness in others. They use fear, pride, anger, lust, and every evil weakness that arises in each individual temporary perspective. Globalist's favorite wedge issues are false fears of climate change, amplified perceptions of social injustice, corporate greed and corruption, nationalism, and Judeo-Christian religions. They demean spirituality, family, parents, moralism, nationalism, capitalism, borders, fossil fuels, and more. And by indoctrinating a purely physical belief of reality, these things pushed from within, can bring a free nation to its knees in the weakness and perversions of endless internal divisions. Reality has both its conscious and physical formula together. And in what these two qualities are now believed to be, they form the state of mankind's current false perspective. One can say a physical reality is all there is; but all purely physical models only vaguely describe what is physically sensed beyond our free-will interactions. In truth, reality is a conscious question, whether you see it or not. Knowing what consciously has, and is now occurring, is required to understand reality. Yet knowing what always physically occurs, is all that is needed for useful technology to advance. And materialists use the advances in technology to claim the political and scientific credibility they need to indoctrinate our children with false theories. Many think that materialist have a superior understanding of reality, when it is so inferior. And they spend that false credibility to get the masses to believe reality has no need for GOD. We hear their deceitful words, labeling Godly opposition to their false science, as science deniers. What do these vastly expanded blurred liberal terms even mean anymore. Back in grade school, websters dictionary defined a racist, as one who *believes* in the racial difference put forth in Darwin's theory of evolution, saying that race reveals lesser to more evolved human versions. Yes, atheist evolutionist Charles Darwin, said by skeletal, muscle, skin, and facial features, and levels of cultural advancement; he saw superior evolutionary advancement in some races. Yes, liberals were closeted racist back then also. And back then, a bigot was defined as one who did not just disagree with another's choice and culture; but exhibited high levels of public physical and verbal hateful attacks towards others for their customs and beliefs; again, like we still see from the left. The Globalists liberal elites, took the strong narrow firm meanings of racism and bigotry that firmly applied to them, and redefined and expanded them in ways to now vaguely and broadly define their opposition. They have made these terms so far reaching and unspecific, that an accusation can be hurled at anyone. The terms bigot and racist have been redefined by liberals to mean any person of Jewish and Caucasian Judeo-Christian Culture; and every word spoken to defend or preserve the good qualities of their culture. Liberals demean all who oppose them using any and every false bullying accusation. They believe destroying lives, or families, or property, to advance their cause, is noble and necessary. All Globalist, Communist, Marxist, Fascist, propagandists will say *I*, in this book, defend racism, bigotry, and all evil things of the past. That being pure nonsense to anyone not under a liberal zombie contrived hatred. A liberal mass psychosis is being propagated through all social structures. We see cultures across geography, language, religion, and race. And each soul, who attains a comfort and security their culture affords, will act to preserve that stability. Preservation is a basic attribute of every culture on earth. But the liberal says if a majority religion and culture in their acquired comfort, in anyway disagrees or denies that every other culture or belief is equal to, and as good as theirs; they are a racist and a bigot. Yes, just by disagreeing. But when the left *always* disagrees with *us*, they say it is *still* the conservative morally religious who are the racist bigots. For *disagreeing* with them, we are racist and bigots; for their *disagreeing* with us; again, we are racist and bigots. Don't waste your time engaging their insanity. Mankind are fragile conscious creatures, and most live in their own individual fractured uncertain reality. We are in a conscious battle between believing and trusting in GOD, or believing and trusting in insanity for continuity. If I hold to *my* own beliefs, if I cling to *my* own ways, I am an individual. Eve transferred her trust in GOD, to trust in herself. Then lured Adam to trust in her, and himself. Eve, deep in her soul, thought that a piece of fruit would make her more-wise, and like GOD, and elevate her up along-side GOD. She revealed her desire towards not wanting and allowing GOD to be GOD. Adam then did the same. No Created Being, can sustain themselves, ever; yet most now believe and trust only themselves, this world, and mankind. This is the essence of the fallen proclivity all souls must overcome. We cannot serve two masters; we must trust *only* GOD'S ways and beliefs of eternal life, or we are in rebellion to the same. Every weak fallen soul must cling to a common consensus for some measure of belief and assurance, even if it is untrue. This is required to ground each lost soul in at minimum a temporary false understanding. This, with the physical distractions of family, friends, health, work, and entertainment, allows many their temporary delusions of an acceptable normalcy of life. If their outer false ritual belief of reality is broken, their false peace and sanity are broken. Disprove their created false surface truth, and a soul must awaken to their reality of not knowing what is true or real. We see many cultures, and the souls raised within them need their established cultural norms. Between all cultures, we see their commonalities, and differences; but coming together too rapidly through mass migration, agitates and destabilizes *all* cultures. The United States has been called the melting pot, being a great social experiment. But what has been proven in this experiment above all else, is that disparate cultures blend very slowly and violently, or not at all, having a strong contrary core resistance. And if they can blend, it is only after a few generations, that a strong blended culture appears. But even in this, there has never existed a balanced blend between cultures coming together. We see countries with strong successful cultures, and less successful cultures. We see the poor mass-migrate to prosperous countries, and the natives find their ability to provide for and teach their own families is diminished and compromised in having to help the less-successful cultures. People never en-masse leave prosperous cultures for poor unsuccessful cultures. And in mass migration, the successful cultures poorer and middle-class are hurt the most; so, they push back against mass immigration. These people are not racist bigots, they are fighting for *their* chance at a good life, and for their own proven culture, language, religion, and way of life. And I tell you, we see GOD does the same. GOD has his culture, and Satan has a contrary culture that is not welcomed by GOD. The Globalist conscious formula cannot abide this; they must as if being a god above us, build a chaotic one world order, void of moral Godly belief and life. Globalist are ardent antichrist souls, conspiring to control all corporations, agencies, media outlets, all sciences, and education centers. They have filled much of these, with their minions who lie, disparage, and shame anyone in their way. They will come for all religious and military institutions next. Migrants flee failing cultures, and for many reasons, yet do not see their culture as the cause for the things they flee. These want financial increase, but rarely desire to flee their cultural, so, they bring it with them. Globalist offer migrants monetary promises that the native culture must pay. Too much, too fast, immigrations from failed cultures brings Socialism, Communism, Anarchism, criminality, sharia, and anti-Christian views. Mass cultural migration stay together to propagate their sub-cultural, as this is who they are. In their large numbers, these weakening the local, state, and national good of the host country. They increase debt, strain budgets, decrease wages, and raise housing costs and unemployment. It is not evil for the *natives* to protect their own successful, safe, culture and religion; yet liberals say it is evil to regulate migration from other cultures and religions. Globalist want the destruction, disruption, instability, and destabilization, that come from mass migration, so they hate and deny borders. But they only say this about wealthy western Christian countries. Globalist, *always* align with illegal migrant minority cultures, against the stable majority native culture. But they hold a hypocritical view for the former

Native-Americans, over-run by mass migration from Europe. They say those European immigrants stole the land. Native Americans lost their lands by open borders and no form of land ownership, against uncontrolled mass European migration. Globalists call white migration, colonization and theft; and condemn Americans for trying to hang on to what they've built. Social zombie emotions are forms of focused hate, and they are taught to instinctively attack, destroy, and silence any who disagree, even as they can't logically defend their views. These are deep in the evils of hell and death that Satan's Globalist push upon this world. Globalist lie and say citizens, trying to hold onto the *good things* of their maturing culture, want to return to the evils of their past, evils they themselves recognized long ago, and overcame on their own initiative. The more antichrist a soul is, the more Globalist they will be; and visa-versa. True racism is ignorance *born* out of the false Theory of Evolution; and godless liberals are true racists. Racism would die if not for new manufactured racism, and constant fanning of its embers by leftist activist. Racism wants to die, as all ignorance does; so, ignorance must be taught to keep it alive. Liberal Globalist Politicians, Professors, Teachers, and Celebs, with minions of followers;

spend their existence, keeping the evil conscious formulas of divisive ignorant language alive. GOD says, if a soul truly turns from their past evil, and is raised forgiven, they are a new creature, and the old person died. Globalist don't want this to be taught, as it disarms them of their weapon of using the near and distant past to demonize the present, and instill fear of the future. Globalists use Marxist doctrine to control and indoctrinate students, through psychological reflections, deflections, and projections, thus dumbing down, dividing, and demoralizing the people. These true devils even redefine the word demoralize; which historically means to destroy the strong morals that bind people as a country together. By controlling the systems of education and media, they divide the people in every way possible. Per-Marxist teachings, for every evil *they* do in the open, they accuse their opposition of doing that evil in secret. But how do people not see it, how do they fall for such tricks. These tricks are only effective if they control the media and education system, and destroy the same understanding souls need, to see these tricks. Now in something similar to the Dunning-Kruger effect, all created zombies, greatly overestimate their understanding of reality; and through relentless lies they confuse the next generations even more. Globalist can claim to be Conservatives to run for office, giving rise to the political term *"in name only"*. Many elected and appointed souls, appearing as religiously conservative, are Manchurians put in place as moderates for those moments of great power grabs. One recent midnight thumbs-down Globalist power grab comes to mind; do you recall that moment. The world today, is a created conscious instability that has historic levels of insanity, as liberals grow detached from all reality. Liberals are insane destroyers of freedom who can't see how they live in obvious devolving contradictions. A fact is, that only true Godly conservatives, are patriotic in nature. The Etymology of the word Patriot, originates with the Greek or Latin word "Pietas", meaning a soul who loves three things: GOD, neighbor, and country. All three and not one or two. If you did not love GOD first, you could not be seen as "Pietas" by loving neighbor and country alone. Again, the left has altered the meaning of this word patriot, to now describe evil. True Patriots are generous and accepting of *controlled legal* immigration from all races, and cultures coming for good reasons, who have not shown themselves to be contrary enemies of *our* culture. Those who come not for prosperity first, but first for our free Culture of personal and national responsibility, having our strong self-determination under a Constitutional peace, and love for GOD; and lastly, for prosperity. The only other immigration question for true patriots, is how many each year, and from where? The point is, immigrants should want to come here to be like us; not to bleed our country dry, and make our country like the undesirable places they flee from. Souls minus morals or self-worth, feed, steal, and live off of their native country's people's work, till their country and people are poor; why would we want them to come here to do the same. I have no sympathy for anyone who abuses and uses others. Liberal welcome poor crime ridden cultures, yet demonize our culture of Godly self-determined Meritocracy. We see in passing that people, cultures, and even religious doctrines change. And minus un-natural pressure, these all change at a slow peaceful pace. We see civil wars around the world, come by one ideology applying pressure, trying to change another ideology beyond their ability to do so. The Antichrist Globalist don't wait for slow peaceful change, as *undirected* natural change is not their goal. They seek a directed change through fear and disunity, by perpetual engineered crisis. If we do not believe as the liberals, they see us as unjust, and in their way. Globalist do not allow value or honor for each life. They do not care that without choice, love dies and ceases to exist. These devils see the world as one giant ant colony, which they rule. This global formula is pushing the Western culture toward a critical mass of moral and financial implosion. Every western preference for the comfort of our current moral commonality, they call racist bigoted entitlement. The vast majority of European, Asian, African, Middle Eastern, Australian, Japanese, South American, North American, and all other peoples, choose to marry people of their own culture, religion, language, and race. So, either the entire world is *preferential*, according to loving common desire; or they are all according to liberal doctrine, bigoted racist. Even most liberals prefer and marry liberals of the same culture, religion, language, and race; even gay and lesbians prefer and choose gay and lesbians of the same culture, religion, language, and race. Wow, every soul in the whole entire world must be bigoted racist; *or,* people are drawn to the loving ease and comfort of a shared life, provided by the path of greatest cultural and religious familiarity, being a path of least resistance. The path which brings the most stability, peace, prosperity, and continuation to their life. And if two love and choose other than their own culture, language, religion, or race, this is a personal *desire* with no inferiority or superiority in that either. But if *anyone* sees a *superior* good, in purposely marrying outside your own culture, language, religion, sex, or race, just for the sake of dissolving cultural and moral norms. Such souls are truly godless, pointless, social zombies, whose narrow reprobate minds, believe they are morally superior to all who disagree. To force unnatural rapid change, destabilizes the majority, which liberals want. I see superiority in the Righteousness of GOD; and like everyone, I choose from where I draw my emulations to find peace and comfort in my life. By our gifted free-will, I believe that for continuity of life, all cultures, languages, religions, and races around the world must protect their strong enduring ways of life. And I am understanding if they don't want me living there, because I cannot blend in. They know I must propagate what *I* know to be true. I believe I must protect and expand the Messianic Judaean community's ways of peace and love. And if anyone says I can't, I see a devil, pushing the spirit of antichrist. We understand, there will always be something in or about the one true Godly way of life that individuals will claim is judgmental and offensive to them. But if I, as a minority Messianic, are offended by them, I am accused of having white Christian fragility, privilege, and other abusive woke nonsense. I am an adopted Messianic Jew, who according to "Wikipedia" are less than half a million souls world-wide; and made up of all races. Does the world in *any* single way recognize our minority status? No. Why? Does some of us being white males, rule out our minority status for the rest? Does being a white male, deny me any purpose of life? Don't get me wrong, I don't believe in or want any minority status or privilege, I prefer to work for and merit what I have and become, I just make the point it is an inconsistent divisive evil before GOD, to claim minority statistical preference. Never the less, I *am* still offended in my religion and culture all the time by all the liberal bigots who do not know me; yet hate and falsely disparage me. Now if you want to keep racial and cultural hate alive as a political weapon, you must constantly fan the embers and flames. You must always pick off the scabs with your filthy fingers and spit your diseased words into any open noisome wounds to keep your lying infectious rhetoric alive, lest it heal through GOD'S love and understanding. No one can say my culturalism is racism, and not be a liar; and they lie about me relentlessly. The peaceful public allowance of all peaceful religions and culture, can never mean denying the *majority* any single place where the free-will exercise of their religion and culture can thrive. Globalist liberals have the *evil* ability to spend vast amounts of time and resources, claiming to fight racism and poverty, while actually making these problems worse. Liberal elites, know fear is a great agent of control, so, they keep people afraid, by exaggerating, lying, and saying only they can protect them. And not all souls, but many do hear, fear, and join. Liberals now tell white children to hate themselves, and for all other children to hate them, teaching the white race is an evil in the world. At the last, FATHER unleashes an unimaginable campaign of shock and awe to try and shake souls from their great blindness, as he dismantles mankind's false reality right before their eyes. Yet most will arrogantly explain it away through their blind desires. This globalist dark political religion, uses natural and created crises and fear to validate their existence and purpose. Falsely setting themselves up as understanding caring leaders. Liberals, being like Satan, hold a deceptive view of themselves as being enlightened above culture or religion. But we see *their* evil culture of false science, false gods, and communistic humanism. Souls have no different colors, and true followers of CHRIST know that all souls are capable of godliness. GOD deals with souls on a purely spiritual level, not by physical race. It has been a natural understanding throughout history, that strong borders and fences make good neighbors among diverse cultures and religions. Is it truly good to draw lines? GOD drew the borders of Israel and said any who call for, or try to alter his borders will face his rath. GOD gave us his laws; are not laws just lines *not* to be crossed for perpetuating peace and continuity. What about your home? Should others be allowed to cross the borders of your yard and come into your home, un-scrutinized and uninvited? If the Globalists say you have the room and financial means, so you must house and feed immigrants. Will their culture, religion, and numbers then overwhelm and supersede yours? If they grow to exceed your family numbers, does your home then belong to them? Can you then claim minority status in their home to reverse that course of events? That won't be allowed, their evil contrivance and deception works only one way among liberal minds. Is it now seen as evil to be a White Judeo-Christian Messianic in the mind of a liberal? Can you hold up any country, culture, gender, or race, not greatly stained with past blood? No; there are none. Are some cultures more affective and productive at building up society than other? How can it be racist and bigoted to examine and consider such cultural evidence? Did the rise of wealth power and charity of the U.S. around the world, have nothing to do with religious moral responsible compassion and self-reliant self-determined hard work. The U.S. was largely settled by Christian Europeans fleeing religious persecution and tyranny. And then European Governments came over, determined to take and rule this land. But through battles and the hand of GOD, this Country was established by GOD as a Christian Nation. And our Constitutional Government was greatly and largely set up as a Judeo-Christian Nation. Could this greatest most powerful and prosperous country, be just a physical chance phenomenon that seeped up from the ground below our feet? No. America's cultural religious ethics, practices, and blessings, that bind strong families and communities under common desires and ideals of self-governance under GOD, are what built this great nation. At the height of slavery, in the U.S. in the 1860's, around 01.4% of U.S. citizens, including hundreds of freed blacks, held slaves (including white slaves). And the other 98.6% of non-slave holders worked hard under GOD to build. Do they not count? And since slavery was abolished 150 years ago, over 99% of this country's progress and wealth has been built. Did the settlers steal this also? If we allow a flood of *ad*verse dependent souls from failed cultures, to displace America's culture of hard work, self-reliance, and self-governance under GOD, will the U.S. magically maintain its strength, or be torn apart from within? Does the strength of this nation flow *only* from the ground, and not from its family-oriented Judeo-Christian culture. The culture of Venezuela, was not long ago, free, prosperous, and safe; and yes, the dirt and rock of Venezuela's ground, are filled with riches. But since it turned to socialism, its people have joined the world's poor, hungry, and oppressed. The people lost their will and way. Are we each meant to aspire to become something, and do something with our life? Or just cattle to be indoctrinated, herded, and farmed, by a controlling body of central power? Globalist see the downfall of the U.S. as a good thing. Marxist Globalist Liberal Elites, will without thought or hesitation, declare that I and the Judeo-Christian Religion are bigoted, racist, and every other name they think of; but I am a spirit of love. I give every soul the benefit of the doubt until proven otherwise. In the 1950s and 60s, the liberal Marxist in the Demo(n)crat party, found their lying voice, usurping race as a political weapon. Again, I was taught that if you pick at a wound, it can never heal. By the mid 1960's, things were getting better between the races in the middle economic class; not fast, but it was recognizable. Progress in the poor rural communities and big cities was harder, and slower. Cities and rural towns are places where, making a living and getting around is hard. Upward mobility in such places is difficult for everyone. So, we see *why* liberals want to destroy the middle class, where healing first occurs. Yes, racism and bigotry exist, and contribute to disparity. But in the late 1960's, democrats used race to fan the smoldering embers of the past, that through GOD'S love, were trying to heal. As a teen, I listened to Dr, Martin Luther King Jr. and found him honest in trying to illuminate the presence of injustice. But once his movement of change grew so big, it was co-opted by demo(n)crats, who killed him, and turned his followers towards hate and violence, opening wide the old wounds Dr. King sought to heal through GOD. New strains of zombie liberalsm have grown so strong, that I see no worldly cure. In past Supreme Court arguments, it was successfully argued that minorities did not have the numbers to systemically disenfranchise the white majority in any unconstitutional way. But minorities today have found and used ways to systemically disenfranchise the white majority, through unconstitutional DEI mandates. This psychological pandemic can only be recognized and called out in the hope of warning those not deeply infected. How has "Nationalist" become a slur; the earth's entire history is born of and exists in Nationalism. Only in the last 1¼ century has Globalism even existed. And most don't know that being a Globalist, means denouncing all countries, including your own. If you want your country to continue, you're a Nationalist, if you want all countries to disappear, you are a Globalist. Now on the Liberal Globalist side, who see no great value in life, they desire to eliminate all Conservative Nationalist. The term "Conservative" means souls who *conserve* and *preserve* their country. "Liberal" means allowance, meaning those who will allow endless change to this country. A Nationalist, says we disagree with liberals on the direction we need to take for our country, but Liberals say we hate anyone who disagrees with us, and we will cut them off from society.

Religious Nationalist Conservatives, filter their emotions through logic and understanding. Liberals filter all logic and understanding through their narrowly-focused emotions. Conservatives filter what people say in a moment, through what they see them do in their entire life. Liberals define what they see other people do in their entire life through what they might say in one single moment. Being a true Religious Conservative, means I see life through GOD'S perspective, and not mine or anyone's. So, I can no longer communicate with liberals in any sense. Unrepentant liberals should not bother to read my books; as they will mis-represent what I say through their liberal biased perspective. We know people far more accurately display their true beliefs through what they do, than what they say. Politically correct speech is a liberal weapon used against conservatives, as mis-spoken liberals are forgiven by the liberal media with their simple statement of having mis-spoke. If in the course of life, I see someone live a certain way, and they then say something contrary to who I know them to be, I must either believe they misspoke, or that they have been living a life-long lie, and have slipped up to reveal their true nature. Words are easily twisted by dark hateful souls, and everyone says things that dark hateful souls can twist to imply something untrue. Which is a heinous act according to GOD. A person who lives by this sword must die by this sword. These will chew their own arm till *their* sword falls. Prov. 21:6, "*THE GETTING OF TREASURES BY A LYING TONGUE IS A VANITY TOSSED TO AND FRO OF THEM THAT SEEK DEATH.*" Psalms 52:1-3, "*WHY BOASTEST THOU THYSELF IN MISCHIEF, O MIGHTY MAN? THE GOODNESS OF GOD ENDURETH CONTINUALLY. THY TONGUE DEVISETH MISCHIEFS; LIKE A SHARP RAZOR, WORKING DECEITFULLY. THOU LOVEST EVIL MORE THAN GOOD; AND LYING RATHER THAN TO SPEAK RIGHTEOUSNESS.*" Psalms 52:4, "*THOU LOVEST ALL DEVOURING WORDS, O THOU DECEITFUL TONGUE.*" Psalms 64:3, "*WHO WHET THEIR TONGUE LIKE A SWORD, AND BEND THEIR BOWS TO SHOOT THEIR ARROWS, EVEN BITTER WORDS:*" James 3:4-6, "*BEHOLD ALSO THE SHIPS, WHICH THOUGH THEY BE SO GREAT, AND ARE DRIVEN OF FIERCE WINDS, YET ARE THEY TURNED ABOUT WITH A VERY SMALL HELM, WHITHERSOEVER THE GOVERNOR LISTETH. EVEN SO THE TONGUE IS A LITTLE MEMBER, AND*

BOASTETH GREAT THINGS. BEHOLD, HOW GREAT A MATTER A LITTLE FIRE KINDLETH! AND THE TONGUE IS A FIRE, A WORLD OF INIQUITY: SO IS THE TONGUE AMONG OUR MEMBERS, THAT IT DEFILETH THE WHOLE BODY, AND SETTETH ON FIRE THE COURSE OF NATURE; AND IT IS SET ON FIRE OF HELL." And don't forget the 9th Commandment. And GOD says, by the measure with which one falsely accuse others, one shall be judged. By Satan's deception and minions of darkness, GOD'S words become a sore subject to discuss for liberals who hate all it says. The Bible is weakened in new versions to allow a liberal agenda to creep into churches; while GOD'S true version is disparaged and pushed from society. Does your Church use a new version of Scripture? Do you ask why? Do you know what was changed from the original version? Do you know why it was changed? Did GOD change it? No. GOD'S children, will in his coming day, live in his one understanding of belief in the mansions of Heaven; but not so for individuals. Pray for yours and all children, that they learn at a young age, that this physical body soon forgets how to live, and dies. And this fallen world and the fallen agents of lust, pride, envy, and greed, that fill it, are just a vapor the eternal wind of GOD will soon scatter among the cold and the darkness. While all over this fallen world, preachers tell their congregants things like JESUS loves to be here with us. No, that is backwards! JESUS came here to suffer, sacrifice and die. Every second JESUS was in this fallen world, surrounded by mankind's created confused hell, was suffering. JESUS wants us to come to be with him, as he lives perfect in FATHER'S presence, among those who live by FATHER'S eternal truth. Every soul in Heaven knows GOD; and every soul there knows how, why, and when, FATHER created all things. Every soul knows the love and great extent and work GOD has endured and endures to give us our existence and choice. How GOD works to hold us together in eternal love, so we don't drift apart into hell's confused oblivion. A soul in Heaven, could stand beside a soul consciously lost deep in Hell, and never know each other in any way. Every soul in Heaven knows, understands, and embraces perfect eternal love for GOD; and exists to perpetuate this eternal loving wonder and glory without end. This is who GOD is, and this is who all souls with GOD in Heaven are. Let us be as they are; or we are a contrivance of evil that divides all confused individual souls against each other. I hope you see there is not one pure perfect justice to be found or had throughout the entire history of this fallen world's imperfect divided, divisive understanding. Minus GOD'S guidance, being lived by all as the truth of creation, this world can only descend farther into Hell's injustice. Taking from each other, blaming each other, and being the shrinking selfish hateful entities they desire to be. The fatal error of all who rebel against GOD, is in desiring the worldly ways of mankind, called privilege, power and possession. It cannot be found, where privilege, power, or possession, were ever a feature of design in Scripture. These words "privilege, power, and possession" allow souls to deny their own burdens. By possession, privilege and power, souls pass their certain burdens and responsibilities upon others. There is wisdom and truth in bearing our burdens. To put our burden upon another, can never be good. But true charitable Godly desire is good, as in helping a neighbor who is worn and broken by their burden. FATHER'S physical Creation and gifts of conscious awareness, reveal his design for our faithful responsibility. There are no considerations for personal privilege, power, or possession, in GOD'S created design. There is no created privilege in being GOD, but there is GOD'S eternal faithful responsible sustaining care for life and reality. GOD did not create power for mankind over others, only responsibility. And GOD did not create wealth of possession over others, only the responsible stewardship of caring for the common needs of others. Many in positions of fame and fortune, glorify themselves by the great burden they try to reduce for others with monetary gifts. When in fact, they live in ease, comfort, and luxury, as they pay others to bear their life's burdens. GOD consciously formulated his creation to serve his purpose, all that lives and all GOD has made, are made to serve one created purpose. Meaning each soul has their part in GOD'S common goal to accomplish, being our own ordained burden to bear, under GOD'S leadership. The physical pleasure of evil desires felt through the temporary external existence, are still here for yet a little while longer. But the greater spiritual pleasures beyond this world, are what will endure. If a soul's faithful responsibilities are not desired and taken up, they fall into mankind's ways of power, privilege, and possession. Spiritual love and teaching, flow from HOLYSPIRIT into all who share responsibility for all souls having eternal love. And if we desire to be faithful conduits of Spiritual responsibility, yet hold strong desires for this world above our physical needs to exist and propagate, we cannot reach our goal. Worship is our highest ruling form of love we can hold. Our worship must be *only* for GOD, who bear by far, the greatest of all faithful responsible burdens, working to perceive and sustain all complex ordered consistencies of our external existence. Physical life is our required platform to exist, brought in its orders of purpose below GOD. Now it is seen throughout history, that every semblance of *enduring* peace, came by the complete victory of one side, and never by a compromised truce. And the coming uncompromised eternal victory will be FATHER'S, who has all his majestic responsibilities, to which he has never once failed. And just under FATHER, we see CHRIST has great eternal responsibilities; including to once take on the fallen flesh and walk among the evil of mankind, to restore truth, mercy, and to sacrifice, suffer, and die. All to which CHRIST never failed. HOLYSPIRIT has his great leading, teaching, and comforting responsibility for all GOD'S true children, to which he has never once failed. These three alone, are GOD above. And the far lesser Gods below came first in Seven faithful Arch Angels, and the Four

Seraphim, each having their great ordered responsibility in the Holy of Holies. To which they each have never once failed. Next the lesser orders of Cherubim and other lower angels in their created orders, having their purpose and responsibility. To which, first the Anointed Cherub Lucifer, and then one third of the lower angels in Heaven failed; yet two thirds never once failed. Then mankind was brought in their orders of their greatest to the least responsibility; to which, all have failed. Yet through CHRIST'S suffering sacrificial forgiveness, around a hundred million souls out of mankind return and are cleansed, new and adopted through CHRIST. So, from the greatest to the least, we all must fulfill our responsibility, faithfully without fail. Scripture shows FATHER brought a creation to be filled with gifted free-willed souls to exist below him. And by all wisdom, we see the greatest to the least responsible burdens, are connected in one common eternal spherical purpose. It then cannot matter where we are set in a spiraling spherical continuous flowing order. Wherever we are created to be in GOD'S good order, there will be those in a greater responsibility above us, and those we are more greatly responsible for, under us. But now the desire of most, is turned to rule over others in wanting mankind's false orders of power, privilege, and possession; and a gaming desire to rise up in that order. Most now seek to rise in this order, to have the more under them, as Satan did. This has become the driving irrational desire in most of mankind. Yet the conscious formula in the perfect created order and ways of GOD remain, these cannot be done away with. But they can be denied, ignored, and temporarily substituted by free-will individual evil desires, wanting to be something GOD did not create them to be. This is a pure delusion, illusion, and a temporary fantasy of being an individual. It is such an ill-conceived temporary interpretation of life. All individual wealth, is filthy clothing that cannot cover the rebellion underneath. In all eternal understanding, this world lasts for less than the blink of an eye. Yet it is all that exists for those who play a god, defining right and wrong for themselves. This contagious delusion, permeates the Worldly Church in these last days of trouble; and it actually triggers these last days of trouble. Many lay out their outer church clothing to gather together and worship, and put on that day's religious spiritual clothing also; so many souls have a different spirit in their hours of church attendance. But as most go home to change their outer clothing, they change their spiritual clothing also. And for the rest of the week, holiness and righteousness are more uncomfortable, limiting, and embarrassing or offensive in a politically correct world. Most professed believers, see GOD'S time, and their time. Do you feel closer to GOD while in Church. We must feel more and more as one with GOD always and everywhere. We can change our outer garment without offence, but in every moment of our awareness, we are meant to wear CHRIST'S Spiritual Covering in Righteous Holiness. Look at the whole of this fallen physical creation, and see all the physically beneficial relationships. But HOLYSPIRIT shows us how each one in excess, is spiritually detrimental. All that physically lives, must acquire energy and sensory information from external physical sources. And by sin and the fall, all that lives has now become a potential source of, not only energy; but also, the physical pleasures and evil satisfaction of others. We all need loving relationships. And seeing this fact that FATHER once existed alone in his perfect internal dreams and desires, with no external loving relationships to be found, we can understand the type of intimate eternal, spiritual relationship FATHER needs and seeks with his Children. And as we dig deeper into this understanding, we see why our soul is ours eternally, but our physical existence must belong to GOD. So, we see that any physical relationship on our part, can never apply to GOD. GOD does not have any external physical needs to exist, but we certainly do. GOD needs and wants us spiritually, not physically; but we need GOD spiritually *and* physically. This is the bright clear distinction between the vast creative Trinity of GOD above, and lesser creative gods below. JESUS said to the Samarian woman at the well, in John 4:21-24, *"WOMAN, BELIEVE ME, THE HOUR COMETH, WHEN YE SHALL NEITHER IN THIS MOUNTAIN, NOR YET AT JERUSALEM, WORSHIP THE FATHER. YE WORSHIP <u>YE KNOW NOT WHAT</u>: WE KNOW WHAT WE WORSHIP: FOR SALVATION IS OF THE JEWS. BUT THE HOUR COMETH, AND NOW IS, WHEN THE TRUE WORSHIPPERS SHALL WORSHIP THE FATHER IN <u>SPIRIT AND IN TRUTH</u>: FOR THE FATHER SEEKETH SUCH TO WORSHIP HIM. GOD IS A SPIRIT: AND THEY THAT WORSHIP HIM MUST WORSHIP HIM IN <u>SPIRIT AND IN TRUTH</u>.*" Remember the term spirit, refers to our soul's attributes of love and need. And FATHER is a vast immense, extremely forceful soul, or person, with his spirit of love and need. FATHER'S conscious forceful projections are seen in the beauty of all created complex rules, wonders of Geometry, mathematical formulas, and pattern sets, like Mandelbrot, and all others. In all these created forceful patterns, we see each and every point of its entire shape or running set, are patterned-force quanta, having a dynamic complex designed reason purpose and function. The controlling guiding beat, breath, and functions of physical life and all dynamic physicality, comes or came by all patterned points of designed forceful guidance flowing down from FATHER. Fallen individuals, tend to see things and ways of reality and life as starting from a physical singularity, a single particle, a single atom, a single protein, and a single cell. But by the conscious primacy of GOD above, physical reality started from countless singularities, aligned by forceful patterned projected desires. A singularity is one perfect physical quality and quantity. And our eternal spiritual-physical dualism and mutualism, requires a single eternal purpose of eternal love, bearing no imperfections. The spiritual hope of GOD'S love towards all, is great; but it requires the greater part of everything spiritually good and perfect from each of us in return. In this we see GOD wants us, for our own eternal good, to see ourselves as under their responsibility for the ways and things we need to live eternal; as it first was in the perfect garden. We see GOD alone is our omniscient guide and source of eternal life and reality. GOD desires to give us all they are and have; being unimaginable to us. But GOD cannot forever give their all to any, who do not now and forever give all they are back to them. Do not ask why I say "now and forever" remember it is GOD who gave us all we are to start with. Without doubt, it is GOD, who already gave their all to us first. We have never given to ourself any force of life, or awareness, or an environment. We are in every essence an ongoing gift and creative work of lively existence, brought forth out of and by GOD. Once we turn to be spiritually and physically one with GOD'S design, studying their works and words, we see GOD has continued to give and give and give and give... So, in our eternal love, we must be good, and give it all back *now, and* forever. Do you see that FATHER gave his greatest sacrifice of love in his only begotten SON, as his purest most precious example of Sacrificial love. CHRIST'S Soul, was set at eternal risk for us, in what FATHER asked him to do for us. John 15:13, "*GREATER LOVE HATH NO MAN THAN*

THIS, THAT A MAN LAY DOWN HIS LIFE FOR HIS FRIENDS." CHRIST willingly desired and came here at FATHER'S request to take on the temporary flesh of a man, and be physically vulnerable to the hateful violence towards him that fills the earth. He came so mankind could pierce the heart of GOD. He came to be torn in his loving appearance beyond all recognition of form, by the hatred of mankind for being perfect. Does this not break your heart? And in return for mankind's hateful betrayal of MESSIAH'S suffering and love, FATHER used CHRIST'S love to build a bridge that can span the great gaps mankind has opened between them and GOD. A bridge formed not of steel, but of the strongest love that exists. A bridge built at an unimaginable cost. A bridge we must cross. A bridge built for all who responsibly and faithfully return GOD'S love, to cross over to be one with GOD. A bridge soaring over the vast sea of mankind's unrighteous unholy lust for the creation. Tell all who might be confused and deceived that the pure perfect truth, is that GOD'S Kingdom is not in any slight manner, a Democratic Kingdom. GOD'S created ways in Heaven and Earth, are not debatable. They are born of FATHER'S loving immutable omniscient eternal purpose and promise. If any created soul is unhappy with the defined created ways of GOD, or with their loving place in the orders of GOD'S creation. If any wishes to rise up in or out of that order of things as they were made and brought by GOD to be; I see a soul who is of their Father Satan. Isaiah 14:12-14 (KJV) "*HOW ART THOU FALLEN FROM HEAVEN, O LUCIFER, SON OF THE MORNING! HOW ART THOU CUT DOWN TO THE GROUND, WHICH DIDST WEAKEN THE NATIONS! FOR THOU HAST SAID IN THINE HEART, I WILL ASCEND INTO HEAVEN, I WILL EXALT MY THRONE ABOVE THE STARS OF GOD: I WILL SIT ALSO UPON THE MOUNT OF THE CONGREGATION, IN THE SIDES OF THE NORTH: I WILL ASCEND ABOVE THE HEIGHTS OF THE CLOUDS; I WILL BE LIKE THE MOST HIGH.*" What happens to souls, with prideful desires to rise up to be more, or different than GOD made them to be? What happens to souls who can no longer believe in GOD? A cavern of belief opens as to the value and purpose of life. We see those who know they are eternal souls, see our awareness is a wondrous gift and creation of GOD, and life's purpose is to cherish each moment in eternal love; and here on earth, to propagate and perpetuate that life as GOD has commanded. But those who see mankind as a soulless godless temporary physical chance happenstance, see life as a physical based awareness, to be enjoyed in whatever physical manner brings them the most enjoyment. Such a huge cavern. We must see all things through the eternal perspective of GOD'S created faithful truth and ordered responsibility; never through the physical lies of privilege, power, and possession. There exists no privilege, or personal power, or personal possessions in following GOD, as many falsely believe. There is only the loving will and plan of GOD, working through their and our shared faithful responsibility towards the spiritual physical reality of the Kingdom as a whole. We are not to remove, deny, rearrange, enhance, or devalue, one word of Scripture; which is only done to liberalize and allow ungodly physical desires and sin to be accepted, or to avoid offending the ungodly. You can live to not offend GOD, or the ungodly, but not both; choose. See *all* things in faithful responsibility under GOD, and see no power, privilege, or possession in between. Seeing life as physical, brings the desire to climb up what is seen as a ladder of success leading to greater physical happiness and security. The word *up* has become a relative subjective term in the fallen physical creation.

Mankind cannot define the top of the universe or the bottom, let alone a rising or descending order of anything in between. In not living in GOD'S ways, and in relying on one's own created understanding, one can believe they are climbing *up* towards GOD'S love, joy, and solace; when in truth, they are falling down toward utter loss and destruction. GOD defines his good ways as up, and all individual ways as down. By letting GOD be GOD, this determines our rising or falling in his eternal reality. Strive in an intense loving desire to find and go to GOD, and they will show you, their ways. GOD says, all who follow their own ways, are Satan's Children, who wrestle against one another to descend into deeper darker desires for worldly temporary power, privilege, and possession. Such souls seek temporary experiential physical safety, entertainment, luxury, comfort, and fascination. And in finding wealth, all unwanted burdens are temporarily born for them. Now there is also a *zealous* sin, in a desire to carry a weight and burden of responsibility and stewardship, we are not truly gifted and created to carry. There is a good humble and contrite religious zeal, and there is also an evil religious zeal, born of pride, that must be overcome. So, seek no religious burden, not firmly put upon you by GOD; and have no desire to set down any religious burden you are firmly called to bear. We must let GOD lead us to where he created and needs us to be and belong. In Heaven there can only be our over-whelming loving responsibility for the overall eternal existence of us all being held together in one belief. At the end of our journey of choice, we will each find we are filled with eternal love, or we are not. And if we are found in Heaven, there is no further danger to our love, we are eternally sealed. Scripture says FATHER is the head of the Trinity, and we can only be found to be his Children through the Blood of CHRIST. Paul said, seeing we are one body in CHRIST, if we sin, we bring uncleanness to the whole body of CHRIST. So, we see and must understand that GOD'S omniscient love is only held by those who always live by it. Remember GOD'S omniscience always aligns with eternal love and perfection and brings our oneness with them. Every individual antithetical understanding is an unsustainable confused belief. All this confused destruction, haunts and taunts this temporary reality. Eternal love is born only through eternal merciful decerning desires, and it is sustained, only by GOD'S omniscient ways. All who reject living in the peaceful perfection *of* GOD, at their last, find a *reciprocal* rejection *from* GOD. By rebellion, these remain outside the pure eternal reality in which GOD'S Children thrive. Endless confused desires, move this falling world and all temporary physical lives toward their final individual destiny. FATHER'S omnipresent omnipotent omniscience cannot mean he knows a soul's perfect or imperfect free-will thoughts before they do; as that would make FATHER the conscious forceful cause of everything souls think and do. FATHER, CHRIST, and HOLYSPIRIT, can and do for a great part, avert their eyes from all free-will acts of evil until the Judgment. Isaiah 59:2, *"BUT YOUR INIQUITIES HAVE SEPARATED BETWEEN YOU AND YOUR GOD, AND YOUR SINS HAVE HID HIS FACE FROM YOU, THAT HE WILL NOT HEAR."* Jer. 11:14, *"THEREFORE PRAY NOT THOU FOR THIS PEOPLE, NEITHER LIFT UP A CRY OR PRAYER FOR THEM: FOR I WILL NOT HEAR THEM IN THE TIME THAT THEY CRY UNTO ME FOR THEIR TROUBLE."* Job 35:13, *"SURELY GOD WILL NOT HEAR VANITY, NEITHER WILL THE ALMIGHTY REGARD IT."* FATHER perceived all things physical in perfect layered oscillations, orbits, and courses. GOD entangles and forces their complex creative purpose downward in their projected patterned quanta of forceful desire that guide and sustain physical life. This created body we inhabit and use within a vast physical environment, and our soul's desires for or against GOD; are what is required to bring free-will choice to all souls. And from our first moment of gifted awareness, most souls reject GOD'S desires, and go their own way to encounter other souls experiencing their own chanceful and willful occurrences. In all this, FATHER continues to hold his creation and life, by his now insufficient guiding forces, to temporarily allow and not stop most of what consciously and physically occurs by all free-will souls. Within this process, we see the eternal-expanding moment of GOD, who sees an ever-diminishing return of fewer and fewer Children. Imagine all physical creation as an immensely complex 3D puzzle with more than $10/^{12\circ}$ forcefully designed and aligned particles. And each point, once formed to fit perfectly and dynamically in place, locked in its relative entangled dynamic ways to all other points. In this, we see how the projected unity of GOD'S perfect purpose and ways formed this reality. Our eternal created place or level is where we will experience our highest eternal grandeur, being fully aligned in the forceful flow of GOD'S eternal loving design.

We will always know we exist; but in GOD, we will know why, how, and where we exist, and know eternal love. I find only secure elation in these choices. Many say I surrender who I am to GOD, as they turn to find other meaningless temporary insane ways and reasons to surrender to. We all desire to surrender to become part of something good or evil; and I choose something eternally loving. Through rebellious eyes, most souls see only half of the equation, and interpret it from a wrong individual perspective. They can't see their soul or reality clearly or correctly. They live in a blurred delusion of reality and life. They think because they see fine details in the thinly vailed physical surface of reality, that their understanding is not blurred down deep. Their godless view has no eternally founded understanding, love, or peace. Their endless false truths and beliefs are forever turbulent, devolving, failing, and confused; and at their end, pointless. I can't comprehend any desire to live in a personally created delusion; or find any reason to abandon the truth of reality. Again, into this existence of chance and choice, FATHER divided out and gifted to us all our desirous space, or portion, of our own forceful free-will soul. FATHER has spoken and created his aether, and all means and ways and physical connections and platforms underlying every created reality; everything required for us to form our inner perceived existence below FATHER, without visually seeing him yet. We are given the means of our soul, to consciously forcefully project, create, and experience our own ethereal particle wave perceptions that enable our soul's rich awareness. All within a vast rich firmer particle wave reality, perceived by FATHER. Through experiencing our desired interpretations *of* our perceptions, we form, accept, and live out our created beliefs of reality. In all of this, GOD is far above all things, and all powerful. Yet he does not consciously direct any evil; and does not witness all evil as it occurs. GOD witnesses and judges all evil later, in opening the books of everyone's stored created perceptions on Judgment Day. All the events of our substantive particle wave creations of perception are stored there intact in the aether. The perceptions of sin, that GOD or any Child of GOD will never see again in eternity, are those forgiven by CHRIST. GOD holds the greater loving force to forgive, dissolve, and forget, all the created physical evidence of our past evil perceptions. Or he may put the evidence of our forgiven evil perceptions in the separated aether in the Lake of Fire. Either way, removing them from

GOD'S and our possible remembrance. But in remembrance of sin, the damned are forever beset in a sea of all their evil perceptions, tormented by ever-present tempting images of theirs, and all sin. In our current moment, though we can't remember all we have experienced at once, it is all there within the sphere of our soul. It is all part of who we have experientially grown to become. At out true forgiveness and baptism, GOD partitions our every *past* created evil; and at our sealing, our evil is dissolved. Every stored imperfection within our self is a cancer GOD must remove prior to our living in Heaven, or they would forever haunt, tempt, and shame us. At our first true forgiveness, our sinful perceptions are not yet removed; but we *are* set in a cleansing process as a new clean soul. Upon forgiveness, we are told to stay pure and clean. Heb. 10:16-17, "*THIS IS THE COVENANT THAT I WILL MAKE WITH THEM AFTER THOSE DAYS, SAITH THE LORD, I WILL PUT MY LAWS INTO THEIR HEARTS, AND IN THEIR MINDS WILL I WRITE THEM; AND THEIR SINS AND INIQUITIES WILL I REMEMBER NO MORE*." Though our former perceptions have not yet faded and dissolved, our soul can be spiritually reset to a new blank start, just as when FATHER gifted us our original measure.

Our overcoming all future sin is a process; and our goal. If we stumble back into confusion and sin, there is a new need for forgiveness of that new evil. This is not a gift to be lightly used, let alone abused. To abuse this gift is to trample upon the loving Sacrifice of our Holy Righteous KING. If you stumble, go and ask again in shame, and wait and hope he forgives you again. We clearly see two parts to what the materialist world calls science. Their first part is the sum of experimental data they can perceive, which if *properly* interpreted and understood, can be of good use. The other part is their then biased false interpretations, suppositions, desired understanding, and their first cause miracles they conjure up to found their belief. They can't see that minus their physical miracles and endless suppositions, their beliefs make no sense. The current science being taught to the naïve, is an indoctrination of *false* beliefs, hopes, and faith; and not true science at all. In following their studies at the battered tenured edges of each scientific field, we hear confessions of their own disparities, confusion, and doubts about their biased struggle. They do *not* try to truly understand reality, they only desperately try to define it to fit their desires. We see the religion of politics infest *every* aspect of what souls see and hear daily. And the sciences are falsely held up as being immune to political bias. Interpretations of scientific data, can never be void of bias, as interpretation means bias. They have no founding controlling empirical facts to build upon, to know this consciously deterministic reality. The newest false scientific expressions and beliefs have always led this world in their political beliefs. There is no other aspect in public discourse, more politicized, monetized, and used to control people's beliefs, than the physical sciences; as they claim to be beyond reproach. They try so hard to interpret, reconcile, and align the data to support their vile godless desires. Isaiah 32:5-8, *"FOR THE VILE PERSON WILL SPEAK VILLANY, AND HIS HEART WILL WORK INIQUITY, TO PRACTICE HYPOCRACY AND TO UTTER ERROR AGAINST THE LORD, TO MAKE EMPTY THE SOULS OF THE HUNGRY, AND HE WILL CAUSE THE DRINK OF THE THIRSTY TO FAIL. THE INSTRUMENTS ALSO OF THE CHURL ARE EVIL; HE DEVISETH WICKED DEVICES TO DESTROY THE POOR WITH LYING WORDS, EVEN WHEN THE NEEDY SPEAKETH RIGHT. BUT THE LIBERAL DEVISETH LIBERAL THINGS; AND BY LIBERAL THINGS SHALL HE STAND."*

If a soul declares they do not believe in GOD, or his every word; it is because deep down, they desperately do not want GOD to be there. And once in control, such souls fill all positions and fields of science with souls who also don't want GOD to be there. They work together politically, to suppress all good spiritual wisdom. They publicly express false supposition and conjecture that deny the necessity for GOD'S existence. They take a narrow, limited slice of highly interpretable data, gathered through instruments and devises, and they, the *churl* (corrupt), falsely interpret it through their desire against GOD. The long line of past failed materialist's theories of reality, were all disproven. And their current theories are crumbling before their blurred squinted sight, including their physical theories on our awareness of qualia, belief, and knowledge. So, how do we each know we are real and exist and live. We each exist in our accepted truth, as proof. But to find the one true pure Religion and Science of reality, we must accept and know the creative Dualism of GOD. Secular Science forcefully propagates their current beliefs on reality, as over-and-over we see them replace their prior failed belief. If it suits their narrative, materialists say they know so much; and if needed, they switch to say, please remember we are in our infancy of understanding. Give secularist any ambiguous reason to doubt and remove GOD from all scientific belief, and en-masse, they desperately try to accept it. Surely the perceptions we create, being ethereal manifestations of quantum particle-wave properties, is the *only* possible answer for how we each sense our awareness. Yet, many refuse to believe this, without one good scientific reason for doing so. Do the sciences lead the people? Or do the sciences just give the people what they also want? Why do secularist reject created particle-wave-based perceptions? Because they know, to create or build anything, including ethereal particle-wave perceptions, is called work. And each complex creative physical work, requires a creative force, *and* something substantively less complex with which to build. But all the physical laws, and the four known forces, have no way of ordering or contributing to such a process. So, what is new force, and from where does force come? Forceful Soul Dualism describes this quantum mechanical conscious cause, process, and order, for both the rising wave and its congealing pattern of particle collapse; yet most deny and can't see how they fearfully run and hide from this fact. The Dualism of a soul, generating projected forceful patterns of quantized desire that build particle-wave structures, fits perfectly. And what can be the new source of our initiated three-dimensional creative patterns of quantized force? And what could be the source and essence of all force? Science improperly measures and studies the effects of existing force as best they can. They are blinded and pushed to see force as being part of a special force carrier particle. That is false and still does not begin to define what force actually is, or where it originates. So, they assume many unreasoned things. Each created particle has a defined measure of patterned projected force entangling it, to be part of an even larger forceful projected pattern. Our every tiny projected desire, has its intentioned and purposed patterned properties, qualities, and quantities. Each soul's willful desire to exist and believe in their awareness, is the only observed quality that acts as the cause of creative and or destructive changes within this over-all quantized sphere of reality; limited only by what each size soul can forcefully reach and affect. Our desires are the only driving intention, for our every thought. Our every projected creative force upon reality is in our intended desire to exist. As we project dimensional representative patterns of our forceful desires to exist into the aether filling the reach of the breathing sphere of our souls; so, affecting the environment we occupy and overlap. We then become aware of our created effects, images, sounds, and our other senses woven together. Our created representative images, sounds, and other manifestations, can never be equated as being our awareness. A created sound or image can never be aware of itself. Our forceful creative ability, and our awareness, are two conscious phenomena that exist just before, and just after our created perception itself. This substantive created perception, is how our memories are formed and stored. Our every conscious desire and subsequent created perception, epitomizes our forceful creative soul. This, coupled to our following *awareness of* our created perceptions, epitomizes all creative conscious souls from GOD above, to us below. Consider again, how all the wakeful or sleeping perceptions we create and experience, are effortlessly created within the growing spherical boundaries of our inner conscious reality. These and all conscious moments of perceived reality, come by pure Dualism. And in all the realities of our existence, we find creative and destructive aspects. Reality springs forth from both that which knows, and that which knows nothing, being entangled by desire. We see our own forceful desires in all we paint upon life's canvas as we behold, ponder, and consider our own created perceptions. The material sciences know we are aware we exist, only through our created perceptions. They also know these are second hand, hearsay, streaming perceptions. They know a vast inexplicable gap exist between our physical sensations, and our awareness of them. Science has *never* found a location or any aspect of created perceptions in brain meat. Yet, minus reason, or one shred of physical proof, materialist say we create scaled down, upside down, reversed, 3D non-existent illusions of our environment within the brain. *Then* within the same few pounds of brain meat, they say we inexplicably experience a vast expansive 3D experience, appearing in every exacting grand scale, and appearing as if outside the brain/body. This is physically inexplicable, impossible, and miraculous; and I could never accept such a fantasy of belief, being beyond all sane reason. We sense and create our perceptions of things where they are and occur, and not by physical mysticism within the confines of the brains wet very fatty neural structures. Do you know the brain is 60% fat? We do not create huge detailed multi-sensory perceptions within wet fatty brain meat. Yes, information for our perspective is provided to our soul by these physical neural senses; but our created experience does not occur within wet fatty brain meat. By our soul's varying scope, spectrum, and breathing reach of forceful projection, we create our representative manifestations of all things external to us; including illusions of time. Now I ask again, how does gravity exist in my dream; yet if I desire, I fly in other dreams, defying gravity by desired created perception. And if neurologist say they see the neural activity of an hours long dream experience, occurring within just seconds of brain activity; how can they say any such thing as timely passing of experience, exists outside our perceived moment? And there is the fact that in my most vivid lucid dreams, I cannot find a weakness in all my senses, emotions, and awareness. My remembrance of certain sleep dream experiences, to this day, are detailed and equal to my wakeful memory. We mostly have just vague, confused, sporadic sleep dream awareness. But dreaming in our most vivid lucidity, seeing, hearing, feeling, tasting, and smelling, just as in wakeful reality; how can one say which is which? And how can one say being awake is not just another created dream, when it is? What is real, and what is reality? Can the wet fatty meat of a brain know the difference between imagination and reality? Scriptures say, "GOD who quickeneth the dead, and calleth those things which be not, as though they were" and "walk by faith and not by sight" and "be transformed by the renewing of your mind." These, and many scriptural verses speak to GOD'S desiring, believing, and calling reality into existence. Now let's stop for a moment, and picture being in Heaven. Please imagine all the souls and things around you that would make for a perfect memory, and please hold onto that picture for later. GOD'S creation provides us strong sensed information through bodies, to which our souls are adjoined, yet we have never once existed in any real physical way. We all need to have one spiritual understanding and belief, tying us *and* our eternal reality to GOD. Fallen mankind's beliefs are formed around the stronger sensed external firmer reality. And this convinces them there is a real difference between created sleep perceptions, and created wakeful perceptions. But both sleep and wakeful reality are only known to us through our internal awareness, so why would anyone believe physical complexity came first and on its own; it seems the far, far, less likely and evidenced thing to believe. Physical reality and life *exist* by GOD'S awareness of them. We in our far lesser ways than GOD, consciously create our perceptions as minute, highly restricted, forceful conscious creators. *Again*, Psalms 82:6-7, *"THEY KNOW NOT, NEITHER WILL THEY UNDERSTAND; THEY WALK ON IN DARKNESS: ALL THE FOUNDATIONS OF THE EARTH ARE OUT OF COURSE. I HAVE SAID, YE ARE GODS; AND ALL OF YOU ARE CHILDREN OF THE MOST HIGH. BUT YE SHALL DIE LIKE MEN, AND FALL LIKE ONE OF THE PRINCES.*" And once again, to be a child of parents, means to be brought forth from and as the qualities of our parents; by the parents passing on who and what they are. And our soul was gifted out from FATHER'S soul above, so, we are bicameral children, being a consciously gifted, physically created dual manifestation of FATHER. And we are all at first, an unknowing, unlearned, minute representation of the exact same qualities and inherent attributes of our parent. FATHER said we were gifted as his children, born by him out of his image, emotions, and dualist properties. Our being of FATHER'S forceful soul, and his created embodiment. Again, John 10:34-36, *"JESUS ANSWERED THEM, IS IT NOT WRITTEN IN YOUR LAW, I SAID, YE ARE GODS? IF HE CALLED THEM GODS, UNTO WHOM THE WORD*

OF GOD CAME, AND THE SCRIPTURE CANNOT BE BROKEN; SAY YE OF HIM, WHOM THE FATHER HATH SANCTIFIED, AND SENT INTO THE WORLD, THOU BLASPHEMEST; BECAUSE I SAID, I AM THE SON OF GOD?" Phil. 2:5-8, *"LET THIS MIND BE IN YOU, WHICH WAS ALSO IN CHRIST JESUS: WHO, BEING IN THE FORM OF GOD, THOUGHT IT NOT ROBBERY TO BE EQUAL WITH GOD: BUT MADE HIMSELF OF NO REPUTATION, AND TOOK UPON HIM THE FORM OF A SERVANT, AND WAS MADE IN THE LIKENESS OF MEN: AND BEING FOUND IN FASHION AS A MAN, HE HUMBLED HIMSELF, AND BECAME OBEDIENT UNTO DEATH, EVEN THE DEATH OF THE CROSS."* But please understand that minute gifted portions of forceful creative god-like conscious creatures, capable of forcefully creating small limited particle-wave perceptions; and being GOD over the entire universe, are unimaginably different levels. We are as nothing, compared to the FATHER of all Creation. Scriptures reveal FATHER'S empirical understanding, and define who and what *we* are. In trying to grasp what are GOD'S and our created reality, we must always in all things, consider our creations of streaming perception, *from* GOD'S vast eternal omniscient perspective; only then can all things be understood. GOD can shut down and summon any soul in a temporary body, or translate their body into the quantum realm of Heaven to truly see and hear GOD. This would feel like not being of, or in this world. And afterwards, one might wonder, was that only a very real inner dream reality, or was it the literal sensed awareness of GOD in Heaven? When reading about the Prophets being taken up in the Spirit before GOD, we see GOD often sent them back to earth in a distant location, to affirm it actually happened outside their soul. The lines that most believe divide external reality from dreams, disappear once you realize that every dream is your created perception of inner reality; and every created inner perception of reality, is a dream. If one is sitting and awake, but set adrift in fantasy. Do they still consciously exist in their seat, or do they then and there exist in their awareness of somewhere else. When startled from daydream, people seem to often say sorry, I was somewhere else. Whether aware in sleep, or awake, or in between, we are each a quantum realm soul, and our outward facing spherical surface is covered by our current perceptions. Does our soul's spherical surface belief, face inward or outward? For our belief, it is both simultaneously. Now, if I meet unknown people in a dream, might these truly be other souls, entangled through a shared created perception in some way? Could other quantum realm souls become entangled with me in dreams; can different souls share a vivid dream experience? In the quantum realm, entanglements of souls at a spooky distance, can occur through vast quantum spiritual channels of mediation. In this confused fallen physical world, all such entanglements are a rare variable; but in Heaven, this may be a new very strong ongoing sense. The laws of conscious and physical entanglement, orders and reorders things with our every desired experience. Now we see many degrees and vectors of waring physical attraction and repulsion; and we see the conscious desired belief variables that rule and affect mankind's dual equations in this world. So, in perception, there can be a physical attraction that is over-ruled by our spiritual rejection. In a once vivid lucid dream of mine, if a true quantum conscious realm entanglement did occur between me and the woman in my dream; was the soul of the woman in my dream, in all reality afraid, or offended by my taking her hand, and gently considering the reality of it? All I know is she smiled. We know that for eternal peace and love, all physical complexity must align, just as all our knowing believing conscious spiritual aspects, must also align with GOD. Now there is no such thing as particle annihilation; particles do dissolve apart more and more as they pass up through the eternal realm back into the aether. The basic singularities of GOD'S aether eternally exist; so, they must go and exist somewhere. Again, how can a perfect physical creation, called the New City of Heaven, endure for all eternity. Only by every free-will soul's desired belief of life within it, aligning. In Heaven, all layers of free-will belief, must in their own manner and moments, desire and choose only from among the great many things and ways in GOD'S creation of eternal life. Yes, in GOD'S Garden, there were so many perfect choices for life; and there was just one created choice for death. But the coming realm of the New Heaven, will have no created choice for death. We must remember the our free-will is an innate quality, but choice is a creation and offering of GOD. The answer to our *eternal* loving peaceful conscious and physical equation, comes by GOD'S omniscience, omnipresence, and omnipotence, over his entire Kingdom. Can you see the layers of responsibility, compassion and Holy Righteous justice within the Trinity of

GOD? Do you see throughout Scripture how FATHER most strongly presents as omniscient; and HOLYSPIRIT strongly presents as omnipresent; and KING JESUS presents most often as omnipotent over us, as FATHER placed all power and dominion over all creation at our LORD'S feet. When FATHER created our sensing body and environment, this allows us our created experience of being in, seeing, and hearing, a reality shared with each other. Mankind in their minute ability, are capable of creating an external illusion by wearing 360° programmed gaming head gear. Seeing this, what greater reality is FATHER capable of perceiving to exist around our souls. Our souls need FATHER'S created medium, or aether, we need a connection to a sensing body relay platform, we need adequate representative information to be sensed, and we have our sensing projected forceful desire to exist. As lost souls exercise their free-will desire to interpret incoming information through false foundational beliefs, their interpretations most often affirm their confused *individual* beliefs of understanding. Such a great complex creation, which gives to us our true or false ordered questions upon questions; needing a true or false ordered answer upon answer. And remember only all *true* questions, have a final omniscient answer, being that an eternal loving GOD is consciously and physically there, waiting and wanting to share with us the perfect ways of eternal life found with them alone. It is all the true *or* false questions and answers, born of our desires that lead to, or away from GOD. We ask Godly true questions, and hear GOD true answers; or, we ask false individual questions and hear our own answers; your choice is so important. At our gifted inception of being a minute three-dimensional portion or space of conscious awareness, coinciding with the exact moment of our body's first cell's conception. We began to ever so minutely perceive measures of information to build our complex perspective and belief. But in these early moments of our eternal awareness, *on our own*, we can find no true understanding. If you desire a false speculation to be true, you will see its possibility. And with enough possibilities, you start to see probabilities. And in seeing enough probabilities, they can fit together loosely enough to form vague problematic false models of reality. By adding many small false desired beliefs together, one may see a larger, stronger, still imperfect belief. Any such strengthening effect of semi-aligned false beliefs is not real, yet it can seem probable, till it falls apart. And most souls actually want and even need to live under such partial weak malleable theories. The current theories as to what is reality, exist on pure religious faith and hope. Multiverse, Quantum Gravity, String Theory, Big Bang, Abiogenesis, Evolution, and many more, can never be proven; these are all concepts of faith. These godless theories only grow more problematic, and more discredited. Satan desperately wants you to believe you create perceptions and achieve awareness, through the wet fatty meat of a brain, and deny you are an eternal soul. Satan wants you to believe reality started from a *physical* miracle, and that force has always existed without source, reason, or explanation. Satan wants you to believe physical life started by itself and evolved by chance. Satan wants you to believe the earth is 4.5 billion years old, and the visible universe is over 14 billion years old. But these beliefs, are all destroyed by the fact that we all forcefully create our substantive conscious perceptions. We see how a new lesser threatening term intelligentsia, is used to replace the far more forceful threatening term "Authoritarian". Notice how we are told to obey these *so-called* highly intelligent souls; and that these souls are the leading *authority* in certain areas. Is this authority bestowed from above, or from within a closed circle of a godless authoritarian system? Who defines these "Authorities" in their Authoritarian System? Such titles arise in the self-righteous arrogance of those who want to force all souls in the world to believe as they do; Globalist seek only worldly authority. Do all these so-called leading authorities, see themselves as such? To differing degrees, yes. Many see themselves doing what *they* believe is good for all, whether they all want it or not. They see themselves as good teachers, and think it is okay to be dishonest, if it advances what they think is good for all. They do not trust in choice for those they see as inferior in understanding. If GOD did not exist, these might have as much right as any, to force their beliefs on others; but the fact that GOD does exist, means they do not. These work to turn souls from following GOD, to follow them. But these worldly leaders are blind, feeling the walls as they stumble down a false hallway to destruction. They sense barely a sliver of the physical side of reality, and interpret it through confused godless desires. Do not hear or follow them, for they lead only to where they themselves are going. They do not see or know where they go, but we do. Fewer and fewer souls faithfully search for the fullness of the one pure truth of reality; and by strong worldly desires, most never even begin. Their many physical assumptions are their stumbling stones. How can you ever see to go forward, if you hold to blinding lies. The highest perfect parts of the truth of GOD, and the lowest *im*perfect parts of this world's false beliefs, are all supported by desire; one upon a desire for GOD, the other upon the desires of fallen individuals for a fallen world of flesh. A false foundation, leads to building upon that same false understanding. And it turns out, you can build quite high upon some false foundations if you can overlook paradoxes and endless inconsistencies. But as the inevitable contradicting data mounts, false beliefs must be malleable and take on new forms. You can never hit the one eternal truth of reality by throwing endless darts at a dark vast universe. Resist speculative assumptions, and be taught the truth by GOD alone. Each found facet or quanta of true understanding from GOD, moves you more into their one eternal truth of love and peace; and each believed false physical conclusion, drags you down into a confused dark individual abyss of time, hell, and death. We all sense more than materialists can define or declare; meaning they can never make any declarations that *dis*prove our conscious soul. They believe in countless things they can't see; yet say they can't see our soul in any way, so they don't believe it is there. But they do see the creative and destructive influences of our free-will soul upon reality. Now they can't see force, or dark matter and energy either, but by seeing its influences upon reality alone, they do believe in these. They just can't see or accept, that the soul and force, are in many ways, one in the same. What can their belief say to me, when *I* see so much more. I believe in GOD, just as much as I believe in myself; and I convey only what HOLYSPIRIT has shown me. Consider the power of indoctrination, in that if you divide a thousand infants into two groups, and teach them a completely opposite understanding of each and every thing. Teaching half, up means towards your feet, and down means towards the sky. Teaching half that right is called left and left is right, sweet is called sour and sour is sweet, the word evil means good, and good means evil, the devil is good, and GOD is evil, given no choice of belief. Then that is who they will become. This is indoctrination. Now as these subjects all reach 20 yrs of age, if you then confine them together in a prepared town, they will separate, and go to war within days because they cannot communicate or accept anything about each other. Consider how in theories, experiments, and equations; if you choose to exclude, or can't incorporate all relative factors, such as the conscious observer's effect, quantum entanglement, and so many others. That all such exclusions, falsify your theories and results. The physical sciences take physical considerations, and measure their current values as they are perceived from their slow small paradoxical frame of observational reference. But defining reality is impossible from a too small too slow temporary perspective, while denying one's own Observer's Frame of Refence Paradox and one's own conscious force projection of created perception, the existence of love, one's awareness of purpose, the paradox of eternity, not knowing 94% of all dark physical reality or the vast difference between light and brightness, and denying GOD. To see reality, we must use the true eternal perspective that comes from stepping outside the whole of creation, and connect to GOD'S awareness of it all. To see it is all entangled and brought together through the dual aspects of GOD. Only a Godly, pure Soul Dualist can ask and know where force originates; and which came first, force or energy. In every way you look at it, in every case that can be made, force must come first. Force moves and aligns all we see. Yet in every case and every way, it further says that forces essence must also come from some initiating source of generation. Knowing this basic order of created reality is so important. Fallen mankind sees force is within and upon everything, but not from where, when, and how it originated. This chicken or the egg question, brings us right back to the fact that *our* perceived existence, could only be created by our own *new* forceful projected free-will desires to exist within our soul. Materialist cannot see that without a desired design of projected pattered purposeful force, all pure physicality, must remain random and lifeless. We exist to experience life through our forceful desires; and all things then become true or false to us through our built up desired chosen perspective and belief. And at our core foundation it is all formed, and reflective, of how we each feel towards that one highest greatest factor, being GOD. A huge unjust wall has been built between conscious reality and physical reality, when none exists. The less than purely physical terms of, philosophy, ideology, and metaphysics, do allow for soul's as forceful conscious necessities, where physics does not. Materialist deny that forceful awareness, and all physicality, have always woven together to form all realities. There can never be a respectable physical theory, seeing they can only sense and consider 6% of physical reality. All their impossible theories, are the confused beliefs of self-appointed Authoritarians, having no empirical proof, or one single non-anecdotal substantiating observation. All we see are their godless hopes and dreams, being far beyond their physical perception, and any testing. They constantly feed on endless sermons of supposition to keep their false failing belief alive. Now is GOD authoritarian? Absolutely. But GOD is the only true rightful authoritarian that exists; we must honor GOD supremely. But instead, we see many levels of dishonor, and lesser honor. Respect is a level of honor, where one admires a soul's goodness. Then there is a higher level of obedience, as one honors and fears a powerful soul's good command. But above all levels, we see pure true emulation, this is where you have respect, obedience, and eternal *love* for that soul. Now it is telling that when science cannot find a way forward, they just roll out new far-flung suppositions that form other unreachable goals to drag the world towards. If you do not believe in GOD, this may seem an innocuous normal thing to you; but once you know GOD is there, you see that is sinister. Their Multiverse, Big Bang, Abiogenesis, Evolution of Species by Natural Selection,

Physical Consciousness, Physical Cell Mitosis, and much more, are all impossible sinister faiths that support and propagate their godless hopes and desires. Can't you see how these physical fantasies are used by these authoritarians to spread their false godless religion. Their illicit dreams never lead anyone anywhere real; yet souls who are trapped by physical desires, need such false religions. As Materialists dreams run far behind and astray of reality, their religious faith, hope, and belief in their illusions, keeps them on a path, where every accepted supposition, births at least two greater, further from reach questions. They walk a desired fruitless path, where all final answers just forever grow more beyond their reach; because their answers don't exist. But again, this is exactly what they want; these souls do not want any final answer. A final answer brings a profound absolute, and their total accountability to it; these souls *need* to exist in their ever-changing unaccountability to any final understanding, in being an ever-changing god of self. This is why the fallen need for time to exist, and not eternity. These are the only conscious constructs, where rebellious freedom to do and be as they believe, can exist. But they cannot see or admit this irrational motivation lies deep within them. In my congealed understanding, their great paradox says if Materialism were right, why does its proof hide and elude being found. The empirical proof of all things being purely physical is just too allusive. It acts as if aware of their search, as it hides from them. Moving from dark corner to dark corner, always more and more steps ahead of them. In their desired search, they have found so many paradoxes, ambiguities, unpredictable constants, universal inconsistencies, and much more; which then requires their faith in hidden variables, perturbations, anomalies, unreasoned fluctuations, physical miracles, and any other conjuring that allows them to hope and continue in faith. In bringing GOD'S model to this point and date, we have experienced not one of these problems! Materialist see only mystery mixed with desire, birthing endless confused un-provable theories. In all their roiling uncertainty, their materialist consensus, stands firm in denying the undeniable conscious primacy of reality seen everywhere in everything. Why do they demand and cry that there can be *no* founding conscious considerations in public school teaching? When it is a demonstratable scientific fact, that much of what they teach is just a physical religion. Dualist see that the forceful conscious side, always leads in the dance of desired creation.

Physical science, flails in great uncertainty in its every appearance. It is like every other incomplete false religion and futile political ideology. Their beliefs and practices are all leaps of faith, hope, and deep desire they can't see, while denying the true reality. This is the definition and condition of all the lost and damned. Eternal souls require something with meaning, reason, and purpose to believe, in order to exist in their comfort of thought and experience. Most souls, seeing GOD'S truth as *un*desirable, it is then *un*acceptable; so, it is *dis*believed to find temporary peace. The sciences see the two sets of laws to physical consistency, forming the two lower realms of reality. But they refuse to see or believe these are both lively realms and platforms of existence. Being the quantum realm of the current Heaven, with the forceful consistency of a great ethereal Kingdom filled with immense ethereal angelic bodies of life, and every conscious soul. And then a more condensed fallen physical realm of firm consistencies, dropped down, filled with its now dying physical life. Both realms exist entangled and overlayed within the spectrum of all consciously created certainties, occupying the same space, with the same center point. Yet the outer boundaries of the quantum realm, reach far greater in over-all size. These realms are two at their quantum divide and connection, yet one in physical purpose. Separate, yet consciously joined at every point across the aethers grid. They coexist under two sets of laws and orders that cannot be unified through a physical perspective without the greater conscious variable. Fallen souls can theorize till they die, but they can never prove anything. Though consciously joined, we can never unify or separate, large and small, fast and slow, left and right, or up and down. Conceptually, we can only seek a forever finer defining of their difference from their oneness. Any true complete division, between any two entangled relative things in reality, would have to extend down to and through the aether, dividing the medium of reality itself. Again, any omission of GOD and our eternal quantum soul's existence in science, causes all to veer from a perfect course in comprehending reality. Such souls must require unseen-variables, physical-miracles, ambiguities, perturbations, unexplained fluctuations, suppositions, tricks of the light, smoke and mirrors, and great false hope and faith to support their fallen beliefs. Existence, or reality, are terms now bent and broken by the endless differing theories of them. Physical science is greatly enamored by any slight alluring beauty and elegance seen in any proffered solution to any problem; and beyond all doubt, GOD'S created truth of reality is beautiful and elegant in the eyes of his Children. But in this world, all beauty and elegance are subjective variables, seen through the eyes of individual beholders. How do Materialist claim a patent on defining created beauty and elegance. One true constant, is that a soul can never say in *self*-certainty that something exists, until it is consciously interpreted and perceived by them to exist in a subjective and objective way. The one true objective way for us, is GOD'S subjective way; so, the only true understanding of reality, is defined by both sides of the coin. One side is conscious projected patterned forceful desired and sensed experiential awareness. The other side is the substantive real particle-wave perceptions we create from the possibilities in GOD'S perceived aether. There is no other way to grasp our own consciously created perceived particle-wave-based existence without both sides coming together. Mankind's sensing soul of existence, exists out from the conscious center of reality, detecting the slow small firm aspects of the physical creation of GOD. Our soul's size, and the size of the physical bodies we are adjoined to, exist near a now fluctuating center, in the size spectrum of all physical reality. A bigger soul, set to a small body, restricts the reach and spectrum of what that soul can physically sense and perceive. Such things are predicted in Soul Dualism, and they confirm pure Dualism. We own no body, but *our* soul needs a physical interface and platform to relay externally sensed physical information to enable our soul's created existence, deep within FATHER'S Creation. Through a sensing relaying body, the energetic information of GOD physical creation is relayed and realized by our soul, enabling us our created certainty that we do exist within this reality. Dualism is the only basis and structure that explains how we can create our second-hand internal reality, within the first-hand external reality, created by GOD. Our created awareness reveals our aggressive godlike creative soul. Our perceptions can only be described as true created spherical particle certainties, which we indwell, sense, and are aware of. No understanding, belief, or awareness, exists within the quantum particles we create to form our perceptions. Nor in any waves or consciously generated patterned force. Science has weaved endless forms and strands of complex elemental compounds together, yet they have never taken any first step towards creating any form of self-awareness. Now there are forced physical qualities and quantities set by FATHER, and at *every* occurrence throughout all life's complexities, we see his omniscient understanding of forceful guidance are seen and required at every step. And we see the size of each soul, determines the size and scope of understanding each soul can grasp. Look at the lesser aspects of life; do they comprehend and seek GOD? No, they do not and cannot. Do not let your soul shrink and sink down and be lost to the physical. Science denies the vast distinctions between the animate and inanimate. Please see that complex chemical reactions and electromagnetism alone could never create, cause, or control the *willful* animation of physical life. Do we go where, and do what the body wants; or were and what our soul wants. All reality is driven by awareness, and a motivated purpose towards life. By this being seen everywhere in everything, comes a great question; how can any purely physical aspect, ever have a knowing purpose in order to sustain the complex countless orchestrations of life. In all life, we see physical interactions everywhere, yet we also see a greater entangled guiding purpose and reason. Physical interaction, animation, and motive, only occur after a forceful conscious desired purpose and reason are projected to create and cause them.

From where is born all purpose, reason, and direction? How does awareness of purpose, reason, or direction start? Why would any soul's awareness stop? Can it stop, or does it move passed its physical losses, towards its eternal choice? Is all desire, purpose, reason, and complex orchestration, born consciously, or is it born within all things physical? How can inanimate physical things, become animated, and then desire certain other physical things? They can't, only souls desire. We all accept that desire is born of the soul, or the fatty meat of a brain, and every soul believes one or the other. Why is there no possible third option? So, for each soul's required belief in something, we all desire to see this from one way or the other. Each soul's core desire towards this choice, enables their basic assumptions in all matters of reality. A frustrating conscious paradox in proselytizing, says that it is never what *we* believe and say, but what another chooses to hear and believe. It is so perplexing that *our* perfect meaning and understanding are most often lost, the moment we speak GOD'S truth and love into the soul of another, it is misinterpreted through their contrary desires. Consider *again*, the undeniable fact, that if you step by step, one by one, deny and remove all the observed conscious forceful considerations from the equation of reality, GOD can disappear before your eyes more and more at each step. But if you then, step by step, one by one, add back and accept all those same observed purely conscious considerations, the conscious desire of GOD, required to perceive the universe, appears more and more at each step. This qualitative and quantitative consideration is immutable; and all that is still observable and unobservable, knowable and unknowable, testable and untestable, measurable and unmeasurable to fallen mankind, are part of the complete spherical equation describing reality. Most souls see and desire and accept only the physical evidence found across a sliver of reality's fullness. So, to one far side, the fewer accept the primacy of the conscious forceful soul, and to the other side, the many accept the primacy of the material aspect. We see an eternal spherical unfolding of all cause and effect, not a linear unfolding; and projected forceful desire of *conscious* reason and purpose, being the true *first* cause. Materialist declare without any proof, that love and all desires, are just chemical reactions. Such a hopeless pointless picture they paint, even as consciously caused effects, are constantly observed and measured everywhere. Our conscious cause, is to exist, to be aware, to endure, to know and believe, to love and be loved. All evil motivational desires are towards power, privilege, and possessions, and are now the cause for the ways of this world. In life, we apply conscious desires, motives, and reasoning to countless things and occurrences observed in our physical existence. We even ask ourselves, why did I do that, or why didn't I do that. This is evidence of our own accepted and denied motives. How can science close the door of conscious force as the cause, seeing this same force influences every sentient occurrence in reality. A materialist perspective is formed by their own many past perceptions of desired interpretation, interlacing their past beliefs to adjust their current belief. We ask no more, than for unbiased jurist to hear our conscious case. Do any hear us? Perhaps one here and there, or might we be pleasantly surprised. In this world there are only conscious acts of physical violence, and conscious acts of physical heroism. Everything mankind the angels *and* GOD have ever done, were conscious motivated acts, guiding every physical outcome. This simple profound Principle, runs throughout every aspect of reality; as it alone, projects and creates all said reality. The aethers one-dimensional physical point singularities, do *not* at any step beyond FATHER'S original aligning resonance, have any additional force or ability to redefine their selves into any rising complexity, or into any perceivable state. Only GOD above, and our conscious souls below, hold the ability to create complex perceivable three-dimensional certainties. Each soul's "Creative Observer Affect" is the epitome of all forceful creative substantive perception, from GOD, down through the ranks. And now all souls work in faithful Godly ways, or their divisive individual ways. Now remember each rising created phase, we call a quantum wave, is actually a soup being formed of patterned force quanta, and countless smaller physical singularities and aspects, far too small for mankind to see in any definition. We call what is too small to sense or see, a one-dimensional aspect, because we fail to comprehend its 3D form. This soup of force quanta and tiny points and lesser unknown particles are so many, and so small, they appear to mankind as ethereal flowing energetic waves, seeing all they can detect are the over-all fields of applied energies. But such waves consist of matter and force. If we know CHRIST, and believe *him*, we see this quantum wave excitation and forceful, so-called collapsing of wave properties in our building far greater particles, is not just a lingering strange after-effect of FATHER creating reality. This is the very essence of "GOD'S Creative Being". The "Quantum Observers Affect" is our evidence of GOD'S and our creative conscious force. Mankind looks into the quantum realm at the particles they forcefully collapse, to form their Standard Model. In all waves of singularities and particles, we see the building phase of GOD'S and our projected conscious creations. From our soul's beginning, throughout our eternal existence, our awareness arises by indwelling our created interpreted perceptions. A spherical Godly eternal soul forever looks outward towards GOD, as our awareness looks out beyond, and exists at, and on our surface. If it did not, we would be stagnant, and not grow. Our reality, exists only in our creating our perceivable 3D physical representations of GOD'S creation. Now one may think our conscious awareness, still exists at a minute level in dreamless sleep; but there is zero perceived evidence to affirm any such thinking. And Secularist also see individual evolving conscious preferences, and an accepted consensus. But we see every minute created belief of desire, as a defined created closed system. And each closed projected system of information is symmetrical to the desire that projected it. So, each created perception is a reasoned projection of a *closed* system of reflective choice. But in dissecting this choice, we see layers and levels of lesser individual desires, meaning an over-all created desired belief. But *we* all are meant to be set under, and follow FATHER'S supreme omniscience, in becoming one with the mind of CHRIST. FATHER'S initial Creation and Garden was a vast perfect complex complete omniscient system. He then gifted and placed free-will souls within it, and put one tiny choice for an antithetical system of belief at the center of it all. Love does not exist without free-will, and free-will does not exist without a true choice. But true choice does not exist without two powerful choices; and so, temptation enters, and our temptation creates choice. Temptation is not sin; temptation is in seeing, evaluating, and knowing the choice set before you. To be tempted by any true choice, you must know and consider the essence of each side of the choice set before you. Now in our growing ongoing process of becoming one with, and walking in the mind of CHRIST, we must learn, and stumble less and less. Imagine a bright glowing path extending out before you, and as long as you focus on the path, you remain on the path. But imagine as you walk along, you hear and see things off to one side and then the other. Things bright, loud, exciting, and tempting. And before you realize it, you forgot the path, and you stepped off and did something contrary to the perfect ways of GOD and you have sinned. So, how are you forgiven again? Do you ask, and JESUS flips a switch, like you kicked a circuit breaker? No. You must feel your shame and repent your betrayal; you must get back on the path, and walk and wait for CHRIST to once again forgive and cleanse you, and for HOLYSPIRIT to reconnect. No matter how long this takes, just get back and stay on the path. And once you sense and know GOD has returned to you, and you then stumble again, and you probably will; it will be more shameful and harder for you to ask for forgiveness for another betrayal; and harder for GOD to return to you. How many times can we stumble and return? I do not know; we can't know how many times Judas stumbled, before his final great betrayal, but we know it was too many. And the last betrayal carried such shame, he could not even ask; and probably could not have been forgiven at that point. This basic "Created System of Love and Choice Principle" applies to all souls under GOD, down to the least of us. This all declares the supreme rule of GOD over all shared reality. Every rebellious choice, is forcefully projected into GOD'S vast created system, and adds entropy, division, and conflict. But we see as entropy finally won, it plateaued, and with a shrinking reality, its affects must also decrease up until all final dissolution is brought by FATHER. In the beginning of what and where we all now are as souls and bodies, there was the projected perfect mathematical desire found in the grand perception of GOD'S universe of life in both realms. That which all souls of angels and mankind were set within to learn. But when Angels and Mankind began to desire, become, and project, contrary patterns of desired belief, GOD'S perfect created perception of belief and desire began its fall, and it has been assaulted and weakened ever since. Again, what if you, from your soul, divided out many tiny free-will, force projecting souls to live within, and share in your desired created most vivid dream perception. But most rejected your desires, and projected multitudes of contrary patterns of desire into your creation; your created perception itself, and the fulfillment of your desired life with them as your eternal family, would become corrupted, confused, and ruined as to your desires. But in you being vastly more forceful, and immense, you could save the fewer souls along the deteriorating course of events who you find do share your loving ways and desires; and perceive a New Heaven and Earth. This universe, like it or not, was created in this way, through the conscious free-will projected desires of FATHER. The physical effects of our sinful *desires*, are all illogical outcomes. It is undeniable, that for forceful projected created perceptions of many souls to remain eternally together in love and peace; this requires one shared truth and belief and way of life between all of us. We cannot circumvent the truth, that all physical bodies and the environment are perceived by and belong to FATHER, for his desires, and not ours. Our souls are aligned to a physical platform from within the quantum realm, yet this fact alone cannot make us love GOD. I find it so strange that when people feel the need to tell me they are non-believers; I sense in their soul a spectrum, ranging from a superiority, to a strong indignation; but I should sense a confused shame. We must know the great love, need, reason, and purpose of GOD, so we might know how to better live and love life. We have already been loved by GOD, and GOD must know they are loved by us in the ways they see love. A soul cannot expect GOD to feel loved, by the ways and desires that individual see as being loving. Does Satan see and feel love the same way as GOD? No. Satan sees and feels love in his own fallen, confused, narcissistic, insane, and depraved way. If you love GOD in any ways, GOD says are unfaithful, irresponsible, divisive, and unwanted, that displease and hurt GOD and their work. How can they feel loved. FATHER gifted our souls in his hope of us each wanting and finding happiness in becoming aligned to the created eternal ways of the loving provisions of GOD. GOD has our shared eternal happiness and provision with them, guiding their immutable desires. So, individual preference beyond GOD'S allowance, including fallen consensus desires, and all individual ways of physical life are contrary. They are disunited, and can never build or hold or know or mean anything eternal. Individualism is all confused rebellion, and a rejection of GOD'S created ways. Individualism destroys eternal peace by requiring constant compromise and creating conflict. It forcefully introduces physical entropy, fission, chaos theory, and a conscious and physical butterfly effect, that push GOD away. Unless we want to be trapped in a state of time, hell, and death, we must know love and adopt GOD'S omniscient belief. A true pure eternal love for GOD, has zero desire to continue to sin, knowing forgiveness for any new sin, comes only by the painful suffering bleeding of CHRIST. We desire to never sin, not for an hour, or a day, or a year; but forever. And by this, we are a lesser perfection, under the complete perfection of GOD. All existential beliefs of life in all souls, must align for *eternal* spiritual love and physical peace to exist eternally. Without GOD'S and our desires being as one, we can have *no* eternal perceptions of Heaven. Without GOD'S projected forceful desire, there can be no earth or heaven or applied physics or mathematics to understand. In pure Dualism, if all souls including GOD, stop receiving and projecting conscious force patterns to sense and create our awareness and the physical half of reality, so zero awareness is present. And every soul did this until all existing residual force faded. Would this leave just a sea of pure uncertain singularities, and an eternally frozen state of conscious non-existence for GOD and us, with zero patterned realities and *all* our created memorialized information dissolved and lost, leaving nothing to spark any soul's return to awareness? This is a question for FATHER. Now secular science holds no desire to know GOD, yet to form their many beliefs. They study the remnants and patterns of GOD'S projected force that created the universe; yet have no desire to see or admit its true source and purpose. Romans 1:20-22, "*BECAUSE THAT WHICH MAY BE KNOWN OF GOD IS MANIFEST IN THEM; FOR GOD HATH SHEWED IT UNTO THEM. FOR THE INVISIBLE THINGS OF HIM FROM THE CREATION OF THE WORLD ARE CLEARLY SEEN, BEING UNDERSTOOD BY THE THINGS THAT ARE MADE, EVEN HIS ETERNAL POWER AND GODHEAD; SO THAT THEY ARE WITHOUT EXCUSE: BECAUSE THAT, WHEN THEY KNEW GOD, THEY GLORIFIED HIM NOT AS GOD, NEITHER WERE THANKFUL; BUT, BECAME VAIN IN THEIR IMAGINATIONS, AND THEIR FOOLISH HEART WAS DARKENED.*" This term "force" applies to every soul's varied strength of projected creative or destructive desires. FATHER'S aether, and all reality are made of purely physical, indestructible, and undetectable, singularities that a false Big Bang cannot support. Now in reality, there are created boundaries that all peaceful loving souls do not cross. And for eternal life, souls must allow GOD to define these eternal boundaries. There is only one eternal nature; yet endless temporary natures. We need to grow our eternal faithful responsible love to never hurt GOD or any soul spiritually, or GOD'S physical creation. We *can* eternally love and dwell in the *latitude* of what GOD allows in our going and coming preferences of sensing his perfect creations; but there is no latitude for the beliefs in life itself. Temporary perspective ways, yield temporary outcomes. Temporary perspectives create all the issues dividing eternal oneness. How I wish I could scream so loud that the world would stop and hear. But I know GOD has screamed many times, and the world hears nothing; it laughs in rebellion, saying why should we hear *you*, when we have our own desires and ways. The line between sustaining *or* destroying reality is this; if out of GOD'S created abundance of good choices, you prefer an orange in a moment, and I prefer a pear; this is not divisive. But if GOD says this is the way he has created physical life to work, and you or I say we want to live a different way, this *is* divisive. Do you desire a different temporary way to live? Do you desire to lead others in how to live, when you have zero ability to do so? If GOD gave mankind a set of schematics, that detail all his complete patterns for eternal physical life; they could never implement it. The only formula we control, is whether or not, to align our ways of life with GOD. It is, and will always be GOD'S perceived creation in which we dwell and exist, so, we can desire, learn, and live in GOD'S created ways. Many believe in a pure physical order of life, and a strange conscious emergent. Others believe in a purely conscious universe, with the physical semblance being an illusion, or a projected simulation. Both beliefs being unaccountable to a creator. Strong-emergence means at an unreasoned uncertain point, within a physical chain of events, that inexplicably a new higher complex dynamic quality, or property, springs forth, out from some completely different far-less complex quality or property. A sudden appearance of anything inexplicable, has for millennia been called a miracle of creation. And using a new synonymous term like strong-emergence does not alter anything. The essence of both the forceful soul, and the purely physical points forming the physical body of FATHER and all his creations, have neither ever emerged, but have eternally existed together. Only each desire is new in each forceful projection, aligning the two together as a new created or emerging creation. Most conscious existence is now individual and disassociated from GOD altogether. We can only be one with GOD, by knowing, understanding, agreeing with, and desiring *all* the intimate loving things of GOD; simple belief and recognition of GOD'S existence is not an eternal loving standard. The fact that we consciously desire, generate, and project patterns of force to create the perceptions of our own existence, show that GOD is not just highly probable, GOD is the *necessary* source of all external reality. GOD'S *required* existence is beyond question to me. And before FATHER'S smaller external creations, his size was not known, having nothing external to be compared to. So, he did not appear inconceivably vast, as we now see him compared to us. So, which of these beliefs, requires more faith, hope, and effort? That GOD, or no GOD, explains reality. The inerrant truth and understanding of GOD'S far larger to far smaller, far faster to far slower, model of physical reality, can only be seen and known through the eternal perspective. The truth of reality, not only requires the conscious *and* physical existence of GOD, it also requires all souls who deny the existence of GOD and *their* eternal soul, to forever devolve in the cold dark absence of all GOD had hoped and planned for them. The truth is that as mankind tries to understand reality in their limited ways, their beliefs and theories cannot be realized. A most important question in all reality is, do our desired created perceptions physically exist? We sense and experience the physical and conscious halves of reality, being dualism; but on which side do our created perceptions exist? We all know we can never see, hear, feel, taste, or smell, nothing; we can only see, hear, feel, taste, and smell something physical. So, knowing we are never aware of the actual physical reality, and that we are only aware of our created perceptions of that reality, our perceptions must be something. All godless beliefs require misdirection and godless desire, hope, and faith, at their core. To me, such beliefs are impossible, nonsensical, illogical, and insane. Imagine playing tennis, or running and jumping over obstacles at your youthful limits of forceful awareness. Do you believe the fat wet meaty neurons in your head, contemplate, calculate, estimate, and in great emotional self-awareness, controls the trillions of coordinated adjustments and movements of our 35+ trillion cells in each millisecond, to work together in a fast fluid graceful union. Or has our soul, over our growing moment gained a consciously willed, fluid, over-all control over the whole body loaned to us by GOD. Do you sense the subtleties of your own conscious force of desire over the physical to exist and move. Can you grasp this feeling, and know it does not occur by physical control. In perplexity, we see the devolving delusions among mankind as they try to define their ever-changing dying physical beliefs. They quickly accept a new belief as they bury each disproven belief. They have a graveyard, filled with the ghost of all their former beliefs. What can empirical or empiricism mean, standing in the center of their ever-growing graveyard of former beliefs, in these times of ever-changing meanings of words. What *exactly* is their current empirical belief in a long line of past beliefs? They have only an ever-growing number of theories, that each now require a *nonsensical* metric to define their odds of ever being proven, we call the ordered magnitudes of impossibility. Empiricism says all knowledge lies only in our second-hand, here-say, physically sensed and consciously interpreted data. How can the blind acceptance of such limited physical evidence reveal any truth, when that second-hand here-say evidence, was forcefully created in the perceptions of biased prior beliefs. Let me interpret their perceptions to see if they are right; but that can prove nothing to them. The strength and validity of all empirical facts, varies between all perspectives to at least some degree. So, we must first find and share that one true governing perspective before we can learn anything. Can you, minus a second-hand, hearsay, biased experience, have a purely physical experience to measure or study? I would say show me one; but you cannot. You have your created perceptions of your perspective of something you believe to be real and external to your soul; something you believe you have measured and studied. Do I say the physical existence is not there, and all data is worthless? Of course not. I do say fallen mankind's, individual small slow desired, confused, godless, subjective perspective and interpretation of all that is external, can have no *perfect* value or meaning concerning reality. Material science uses an immense false dart board, minus GOD'S bullseye. We all interpret and believe things within the boundaries of our biased conscious experience. If you desire to, you can easily believe any lie you want; and easily deny you are doing so. You can believe with all your heart you are experiencing the physical world first hand in a pure physical-to-physical way; but at the end, you will find you have deceived yourself. Our experiential existence, is all an internal created indwelled ethereal manifestation. Fallen mankind has never taken one complete and perfect first-hand physical measurement, and lack both the understanding and ability to do so. Surely the external creation is there, but our created awareness of it, is only *our* second-hand created part of our equation that defines *our* reality and existence. Again, we see above all things, mankind's long history and pattern of disproven empirical facts. By all empirical observation, it was science who once said the world was flat. Then science said the sun revolved around the earth. Were those ever-empirical facts? No. And neither are what they believe today. So many dead empirical facts, born of elitist intelligentsia. And who declares who are these elite, well, it is always the elitist themselves. Elitism is term along-side intelligentsia, created to replace the hated term authoritarianism. The purpose of all these terms is to promote a godless scientific superiority of mankind. Satan is the father of elitism, and father of every lying lesser order. All that is declared to be true, yet outside GOD'S one truth of reality, is born of elitism. All experiential observance requires interpretation, for it to have meaning for us. Our desired interpretation of information, is *never* formed or managed by any physical aspect itself; two men look at one woman, one sees more beauty, one sees less. Yes, all received physical information, is exactly what it is; but the fallen individual subjective interpretation of it all, is never an exact representation. Some are close, but at some level their biased interpretations differ.

Do you not see how this obvious pattern has occurred throughout this world's history. See how a new untenable physical theory of reality is eagerly accepted in hopeful desire; and how its list of unseen variables, perturbations, paradoxes, required assumptions, and physical miracles, grow with its study. This process of collapse unfolds each time, and a new fallacy cycle begins. An endless cycle, as mankind hopes in each new theory, till they are convinced again they are wrong. We further see they cannot easily accept they are wrong, they will cling and patch and brace the old, until a new acceptable fallacy is found. It is so clear to us, that only their strong core desires against GOD can perpetuate this ongoing cycle. Mankind is now at the point of seeing great and growing trouble with their current model. I don't see where they can go next, once their current model collapses, and neither do they. They are running out of false road, as the precipice and cliff of their blind eternal insanity approaches. Fallen mankind's driving desire to disprove GOD, guarantees the failure of their every effort in life. Each generation sees themselves as having greater understanding than those before; but as each generation falls and worships technology, they find less true understanding, being more and more ignorant of, and further from GOD. Technology and the truth of reality, are not the same thing. And in this fallen world, these travel in opposite directions; tech grows, as truth shrinks. This world indoctrinates all its unknowing children with absurdities and evil desires, and tries to ban and censor everything that would cause them to question this satanic indoctrination. We want all to know, one true fact, being the physical sciences of this fallen world, have been under godless control for as far back, as back goes. From the beginning they have been corrupted from the inside out. Each new physical theory relies on the same untenable non-conscious physical miracles that failed to sustain all prior theories. They refuse to see their foundational problem, as they choose to just reorder their higher structural problems. Fool me once, shame on me; fool me twice or thrice... The sciences are being forced more and more to ask; could the so-called constants of nature, all be variables. The scientific belief in immutable physical constants, is just one of many errors in their belief. Malleable conscious determinism is the only controlling true constant. The life of godless souls is a relentless hell in trying keep up such a façade of belief, and deny what is true. To hold a belief in physical determinism;

yet to the contrary, see how reality bows down to conscious primacy. This fallen world hangs on fiercely in their ever-devolving self-delusion. If I squint my eyes until everything is very blurred, I can think I see things not actually there. Is this the Materialist method; or does it require closing one eye and tilting one's head sideways. This sounds sarcastic, but it is so hard watching the world run off the cliff of a volcanic caldera with their eyes blinded. There will always be great unknowns, cracks, and voids, in materialist understanding. We, who see reality from above and beyond this slow small physical plane, wonder how souls hold to such unprovable impossible things. With all the neural mapping science has done on the brain, they can't begin to define what a physical based consciousness is, or where it is, or what and where is a brain's created perception. Show us any purely physical process for how and where in wet fatty brain meat our perceptions are formed; and from what, the wet brain meat forms them. Does their neural mapping of a brain's interface activity, reveal any defined physical way or path to create experiential awareness? No. They have found not one thing that shows brain meat's ability to create three-dimensional multi-sensory particle-wave-based perceptions; let alone our following other-worldly experiential indwelled awareness of our perceptions. They hold tight to old ignorant superstitious beliefs; so, I see them, the same way they see me. But I have vast great observable reasons for my belief. They have no reason beyond a desire to disprove the existence of GOD. Every resonant spark in every nerve and cell in the entire body, is part of the sensed and sent information between the body and soul. All back and forth quantum-classical information, is interpreted through our quantum souls desire, allowing us to create and experience a rich mobile existence. Most believers hold a blurred idea of GOD, and think we are incapable, or not meant to know GOD in any understanding; but it is the opposite. An enduring intimate knowing eternal love, is what GOD wants with us. Not any unknowing, confused, distant, weak recognition. Love is the strongest most intimate of experiences we can know. And we cannot love anyone beyond how well we intimately know them. The more we know GOD, the more we can love GOD. JESUS tells us to seek him out and learn of him like a treasure. We can love GOD in our great understanding, or love them in a less-understanding ways. I attest there is no greater treasure in reality than the knowing love GOD shows his Children. I have asked myself, and I ask you to consider deeply, if one can be selfish when it comes to wanting to know GOD. Can someone want or need too much understanding and wisdom from GOD? No. Wisdom is the loving Holy interpretation of knowledge that brings not a worldly; but a shared common Godly understanding. Are we meant to know little of GOD, or are we called to intimately be sons and daughters. We are called to follow, know, and worship FATHER in obedience under CHRIST, and emulate all *his* loving understanding, wisdom, and obedience. Can there be a highest exchange of shared love, minus total shared understanding, reason, and purpose. Without shared understanding and belief, how can there be a shared anything. If GOD sees and says we need to live in his ways that propagate eternal life, love, and peace. Yet many say; according to our own desires, understanding, and beliefs, this does not make sense to us, or please us; who is wrong? In getting truly close, and truly knowing GOD, all things of GOD then make sense to us. So, we must learn, understand, and see, the eternal reasoned purpose in all things pertaining to GOD'S eternal ways, to truly be close to them. Why would FATHER gift us in our creative forceful awareness of reality, being quite capable of knowing and comprehending his great fullness, if not for us to do and experience just that. And why would GOD gift soul's incapable of this; there could be no loving eternal purpose or benefit between FATHER and such a soul. FATHER could never do such a useless thing. Every soul is gifted in the hope they would follow, learn, and emulate, the life perpetuating example of CHRIST'S love for FATHER, in every loving way. We are created to know GOD in our ever-ascending awe of the great richness of their love, under FATHER'S shared understanding and belief. The reach and potential of every loving soul in the quantum realm is infinitely greater than it appears to most. But if filled with illusions of worldly physical desires, most are ensnared in a faith born of those same desires. Why do so many choose and love most, this fallen illusionary unfulfilling physical existence, and not the conscious creator of it. How and why can one ever love a work of art, above the Artisan. Because they know only the creation, and not its creator. How can so many see the awe and vast complex wonder and beauty of the creation; yet not see the far greater beauty of sacrifice, suffering, and work that went into its creation and purpose. The effort hope and love poured out, and invested within it, and its huge loving purpose and reason that seems so inconsequential to most souls. For most, it has become only the moments of physical experience they seek and desire. This is the most horrific blindness that can come upon any soul, being a self-*chosen* self-centered desired blindness, diametrically opposed to our true purpose and possible closeness with that greatest of all Artisans. If one chooses to know and prefer the creation, and not the creator of it, such a soul can never reach the fullness of their ability to know anything. This ability was gifted us in sufficient measure to intimately know GOD. But without knowing GOD, we are never allowed to *truly* know anything else, ever. In knowing GOD intimately, life's purpose changes before our eyes. Scripture says this does not occur within all *casual* professed believers. JESUS said in Matthew 15:8-9, *"THIS PEOPLE DRAWETH NIGH UNTO ME WITH THEIR MOUTH, AND HONOURETH ME WITH THEIR LIPS; BUT THEIR HEART IS FAR FROM ME. BUT IN VAIN THEY DO WORSHIP ME, TEACHING FOR DOCTRINES THE COMMANDMENTS OF MEN.*" A great change from a temporary to eternal perspective of belief and understanding, occurs only within the fewer who escape their desires for, and belief in a worldly based reality. Who among us all, finds true refuge in the spiritual Kingdom of GOD. Matthew 13:19, "*WHEN ANY ONE HEARETH THE WORD OF THE KINGDOM, AND UNDERSTANDETH IT NOT, THEN COMETH THE WICKED ONE, AND CATCHETH AWAY THAT WHICH WAS SOWN IN HIS HEART.*" Matthew 13:44-46, "*AGAIN, THE KINGDOM OF HEAVEN IS LIKE UNTO TREASURE HID IN A FIELD; THE WHICH WHEN A MAN HATH FOUND, HE HIDETH, AND FOR JOY THEREOF GOETH AND SELLETH ALL THAT HE HATH, AND BUYETH THAT FIELD. AGAIN, THE KINGDOM OF HEAVEN IS LIKE UNTO A MERCHANT MAN, SEEKING GOODLY PEARLS: WHO, WHEN HE HAD FOUND ONE PEARL OF GREAT PRICE, WENT AND SOLD ALL THAT HE HAD, AND BOUGHT IT.*" Mathew 13:51, "*JESUS SAITH UNTO THEM, HAVE YE UNDERSTOOD ALL THESE THINGS? THEY SAY UNTO HIM, YEA, LORD.*" The reality of the Kingdom of Heaven, spiritually exists within each Godly soul. This is where Heaven or Hell truly exist in all spiritual realities, right there in our soul's core desires. To experience the Kingdom of GOD, is *not* what surrounds our personage, as we enter its walls. Luke, 17:20-21, "*AND WHEN HE WAS*

DEMANDED OF THE PHARISEES, WHEN THE KINGDOM OF GOD SHOULD COME, HE ANSWERED THEM AND SAID, THE KINGDOM OF GOD COMETH NOT WITH OBSERVATION: NEITHER SHALL THEY SAY, LO HERE! OR, LO THERE! FOR, BEHOLD, THE KINGDOM OF GOD IS WITHIN YOU." Like all things eternal, Heaven is mostly the conscious spiritual aspects of GOD'S loving ways of life, being there *within* the heart and desires and all the created perceptions of its inhabitants. The essence of Heaven is a people being one with GOD'S love, *not* a physical city where souls physically go to find physical joy. Yes, there *will be* an actual physical city called the New Heaven, perceived by FATHER for his SON and us. But a city only becomes Heaven, by the Godly desires and eternal loving ways filling the hearts of all who dwell within. Fill it with all contrary desires, and it will turn to hell, just like this world. Many who profess CHRIST, and attend Church for much of their life, who if allowed in, would say this is not Heaven; as it will not be what anyone physically wants or expects. To most souls, it is family, friends, and an eternal supply of the many things that fill this world, that is their idea of heaven. And again, when most create their first fantasy images of what Heaven might be; they don't realize that GOD is not even in their first images. Is GOD foremost among the images you create of Heaven? If not, it should be. For the fewer, by truly understanding *this* world, *it* lacks the appeal it holds for the many. To the fewer, this world is faint and fading, and filled with divisive ever-increasing evil. It is completely corrupted and disfigured by sin, destined to be dissolved and replaced. The colors of this world are flat and dull. Again, I often say, if a man came holding a mysterious glowing sphere, saying if I hold that sphere, I'll experience a physical elation not known on earth. This I know I must refuse; but most may pay a lot for this temporary experience. *We* know everything temporary has no true or eternal value. Such physical experiences are always followed by the fading of the same, and a desire for more. All that we are given in Heaven, shall forever remain. To have something dear, and have it taken, or to have it, yet the desire for it disappears over your expanding moment; these are the things of death and hell, these are the ways this world offers. At the end of it all, you'll see all you physically received, were illusory addictive pleasures, that were taken back until you lost every physical thing you ever desired, and everything you might have ever had, or been, for all eternity. Everything imperfect is always temporary, and everything temporary is imperfect to GOD. All that occurs, is judged according to the eternal laws, ways, and desires, of the Creator of it all. The pain of losing souls we love, is not in the ways of GOD. And I have no desire to lose those I love. My greatest love is for GOD, and then those of GOD. My parents, children, and family, I love them as I love myself. But I love GOD far above myself, so, I follow and obey GOD far above them. For anything to truly be a good thing to me, it must be found eternal in its created orders; for this is the mind of CHRIST. 1st Cor. 2:6-16 *"HOWBEIT WE SPEAK WISDOM AMONG THEM THAT ARE PERFECT: YET NOT THE WISDOM OF THIS WORLD, NOR OF THE PRINCES OF THIS WORLD, THAT COME TO NOUGHT: BUT WE SPEAK THE WISDOM OF GOD IN A MYSTERY, EVEN THE HIDDEN WISDOM, WHICH GOD ORDAINED BEFORE THE WORLD UNTO OUR GLORY: WHICH NONE OF THE PRINCES OF THIS WORLD KNEW: FOR HAD THEY KNOWN IT, THEY WOULD NOT HAVE CRUCIFIED THE LORD OF GLORY. BUT AS IT IS WRITTEN, EYE HATH NOT SEEN, NOR EAR HEARD, NEITHER HAVE ENTERED INTO THE HEART*

OF <u>MAN</u>, THE THINGS WHICH GOD HATH PREPARED <u>FOR THEM THAT LOVE HIM</u>. BUT GOD <u>HATH REVEALED THEM</u> UNTO <u>US</u> BY HIS SPIRIT: FOR THE SPIRIT SEARCHETH ALL THINGS, YEA, THE <u>DEEP</u> THINGS OF GOD. FOR WHAT MAN KNOWETH THE THINGS OF A MAN, SAVE THE SPIRIT OF MAN WHICH IS IN HIM? EVEN SO THE THINGS OF GOD KNOWETH NO MAN, BUT THE SPIRIT OF GOD. NOW WE HAVE RECEIVED, NOT THE SPIRIT OF THE WORLD, BUT THE SPIRIT WHICH IS OF GOD; THAT WE MIGHT KNOW THE THINGS THAT ARE FREELY GIVEN TO US OF GOD. WHICH THINGS ALSO WE SPEAK, NOT IN THE WORDS WHICH MAN'S WISDOM TEACHETH, BUT WHICH THE HOLY GHOST TEACHETH; COMPARING SPIRITUAL THINGS WITH SPIRITUAL. BUT THE NATURAL MAN RECEIVETH NOT THE THINGS OF THE SPIRIT OF GOD: FOR THEY ARE FOOLISHNESS UNTO HIM: NEITHER CAN HE KNOW THEM, BECAUSE THEY ARE SPIRITUALLY DISCERNED. BUT HE THAT IS SPIRITUAL JUDGETH ALL THINGS, YET HE HIMSELF IS JUDGED OF NO MAN. FOR WHO HATH KNOWN THE MIND OF THE LORD, THAT HE MAY INSTRUCT HIM?

BUT WE HAVE THE MIND OF CHRIST." The gift of being one and perfect in GOD'S beliefs, ways, truth, and wisdom, begins not by sayings, or thoughts, or feelings; but our true gifted Baptism of HOLYSPIRIT. Then we hear, learn, know, and faithfully follow the one pure shared understanding, that underlies the entire purpose of all created reality. There are parts of what HOLYSPIRIT teaches and affirms that are hard to accept, but we must endure education. This is our growth, as worldly institutions are filled with lustful desires to live in shared godless physical beliefs, seeking false comfort among one another in false understandings. But why do most who say *they* believe, refuse to live in the eternal ways and understanding of GOD, and refuse to be one with GOD in reality. It is an attribute among free-will souls, who want a physical Heaven yet don't want GOD'S defined ways of life. By strong desire, these must be the judge and a god who defines the ways of their lives, refusing to see their own hell and death till it's too late. Peter speaks of Paul's writing in 2nd Peter, 3:16, "*AS ALSO IN ALL HIS EPISTLES, SPEAKING IN THEM OF THESE THINGS; IN WHICH ARE SOME THINGS HARD TO BE UNDERSTOOD, WHICH THEY THAT ARE UNLEARNED AND UNSTABLE*

WREST, AS THEY DO ALSO THE OTHER SCRIPTURES, UNTO THEIR OWN DESTRUCTION." This says weak believers misinterpret Scripture, leading them to destruction. At their last, those who desire this vanishing physical creation mankind has destroyed, will forever lose all they love. And in never-ending emotional anguish, they gradually forget what love is, it being of GOD. It will become like a whisper once heard they can no longer remember. It will be a faint unrecognized echo ringing down their inner canyons of endless fading longing for just one tiny moment of stable understanding and belief. But we who long for the Spiritual, who are not of this world, find our hope for eternal knowing love will be fulfilled. The fallen decaying remnant of this world, still holds a weak physical surface beauty and facade here and there, hiding the detestable underworld of decay, a physical beauty which appears richer as one forgets GOD more. Its physical beauty is faded and fading when seen in truth. Every physical aspect of this fallen dying world, seen through the eternal perspective of GOD has lost all *eternal* beauty; and its *temporary* beauty has no eternal value. Why will the physical sciences never declare that the supreme question should *not* be, what is all matter and energy; but what are our inherent eternal forceful conscious qualities that project, create, comprehend, and raise our awareness? How can our complex perceptions passively form themselves? What projected forces, acting upon what physical realities, can orchestrate and form all the information coming together to define our multi-sensory perceptions? This must require a consciously desired intention and purpose towards creating our awareness of self. Secular science cannot acknowledge their profound contradiction, where on one hand saying *all things* sensed are of the material energetic world. Yet as to sensing our created perceptions, being the only things we literally sense; with zero evidence and no more than their desires, they deny our perceptions have any substantive nature. But why? Well, because any ethereal substantive nature in our perceptions, *must* be created through the wave function. And accepting this fact, aligns one to the conscious primacy of knowing all physical realities must be created through the Conscious Observer Effect, declaring Dualism a certainty. Which then defines our soul as a quantum forceful entity, capable of godlike creations by our every projected desire to do so. It's the only possible understanding that can explain the physical qualities of our created perceptions.

And it follows that all existing firmer complex substance, external to us, must be the firmer perception of a far greater projected patterned conscious force. If we open our eyes and mind as wide as the creation, we can see and understand all these forceful logical implications. Every physical quality, and every feeling and thought, has its consciously projected patterns and orders of forceful desire at its foundation. Our soul is a multi-sensory, logical and emotional generator and projector of patterned forceful desire. Again, in pure Soul Dualism, there is no gray area or blurring between conscious and physical primacy. Desire *is* the very first primal element of force, and the first primary aspect of our soul. This does not say, and does not mean we always knowingly and rightly consider all things, before we actually desire something. Sadly, most souls desire all things in foolish ignorance. Our wrongful desires can grow and rise up in force, before we truly learn, know, and choose what we want. It is a process of going from who we were, to who are in each new moment, and who we will eternally be. If you see that this boils down to our core polarity of force, for or against GOD, you have the first sight. But our force is currently changing, even if in unrecognizable ways. Arguing the founding principles of reality, we find every desiring soul is in control of self. And force can never be seen to originate from any other source than each and every soul. And again, all materialist theories concerning memory, come without any defined process. They are simplistic unrealistic beliefs of blind faith and hope, void of reason or purpose. They flail about religiously cheering each other down a fruitless reversed upside-down inside-out topless bottomless path of false temporary logic. Our desires and actions are all the good or evil we are in each moment. Every thought, action, and choice we make here and now, we are truly accountable for them all. You may think to say I have cast my sins upon the LORD. Yes, you can say this, but until JESUS sees a truly *repentant* soul, and chooses to take up his suffering for your sins, they are still yours. We are never forgiven by asking, we are only forgiven when CHRIST actually forgives us. Do you think JESUS has no choice, and is bound to take our sins at our request. Now there was never a moment in reality when creative conscious awareness did not exist. There was never a point in the total spherical moment of reality, when the countless physical singularities of the material half of FATHER, were not there under the authority of his desired forceful creative awareness and belief. Can anyone see our consciousness as a passive element? No, we see it is an aggressive element. Our soul is the ruler and creator of *our* reality, we are not at all passive. As long as our souls are filled with the aether, and connected to a sensing interface that relays information that in its ways and abilities reflects an external reality; we see our created perceptions and believed interpretations of that reality, can never fail to appear. That which knows, forever creates and is aware of that which knows nothing and which was created. Dreamless sleep, is our soul's disconnect from the adequate patterned external physical information, and waking is our reconnection. One moment we exist, the next we do not, then we exist again. FATHER, JESUS, and HOLYSPIRIT, are one in mind as one GOD; yet FATHER is GOD to them also, just as he is to us. The great distinction, other than our soul's size, force, and reach, that delineates us as lesser creative gods below. Is the sealed and set Triune GOD above us, each having eternal connection to every physical point in their three bodies. The Trinity of GOD, are three great, greater, and greatest conscious forces, that act as one, holding all power to provide the *required* physical side of our lively reality. FATHER gifted our souls, and created us physically, having forceful control over our particle perception creation. And the larger gifted souls of angels, have a greater forceful reach, strength, and knowledgeable belief, than we do not. Yet they too, do *not* own the singularities that form *their* ethereal bodies, or their created perceptions; both angels and mankind are given only their soul eternally. As FATHER controls all life's functions through his pre-set complex patterned force quanta, being funneled down to orchestrate every process of life. This line there between *our* sensing, feeling, and moving the body, and GOD'S ongoing work to sustain all its life's functions, mankind calls a conscious and sub-conscious divide. We are only the driver of a leased, now deteriorating body, most souls constantly abuse in ways it was never created for. FATHER is our body's designer, its material, manufacturer, disgruntled mechanic, leasing agent, and even the fuel. Most go about not knowing the constant work of FATHER to keep life going in the body most believe they own. Most even believe it is who and all they are. Can you imagine having awareness and control over every singularity in this body forever, it truly being yours. Then, just as we now desire and control muscles movements at our projected desire, we could alter the shape, size, composition, textures, and appearances of the body. And like GOD above, we would then be eternally self-sustained by our projected forceful desire to exist. Every patterned desire we project, would align our physical body's reality. And yet, we would still be far lesser eternal Gods, still having all our relative limitations under FATHER, in being minutely gifted in what we are. FATHER, by great force, could still confine all that we could desire under greater crushing force, by the desired thought of doing so; just as within the lake of fire that will imprison all evil. Whatever FATHER considers internally, becomes real internally; and whatever FATHER speaks and perceives externally into his aether, becomes real externally. GOD'S shared desired awareness, defines their force, their knowing, and their splendor. Such vast reaching awareness, is staggering for us to comprehend, and it is the reality into which our minute eternal creative awareness has been set in hope of love. So, consider most carefully what you do with your gift. Every particle in the entire universe, as it all exists between FATHER'S hands, does not constitute his physical entirety. The universe exists between his hands, outstretched before his face. Now GOD should not be said to live, GOD is the source of all that lives and exists from every direction, to every direction, and even being every direction. GOD is not an emergent, or alive within existence; and GOD did not evolve. If you falter and go down any such road, you will be lost into Panpsychism or other dead-end paths. It is one GOD alone, who desired, considered, and caused all consistencies and complexities that form all reality. In our knowing that desirous projected awareness has *always* existed, we see an ardent necessity for GOD'S existence. From pure Dualism, we get all the rising qualities of all the created realms, formed by the axiomatic principles of the uncertain possible and probable certainties of entanglement, brought not once by chance, but by rising free-willed conscious determinism. And so, we see all the undeniable answers to every question that most of mankind do not want asked, or answered. To all who deny this, if you can be honest with yourself right now, you will see the comfort of your false belief, rests in creating evermore unanswerable questions and doubts, and in never finding the final truth. Please understand that concerning our choices for, or against GOD, that no matter how hard you try to create an illusion of plausible deniability, accountability to GOD can never cease to exist. We all see the evidence for this, as our choices and beliefs are revealed to be simple matters of strong desire. Now once FATHER has applied certainty of purpose to his inanimate creations. Can we, the weaker conscious force, then weaken or alter his creation? Yes, but only after FATHER gifts and allows our free-will destructive choices. We see in our near proximity, that in controlling a body, we now have our conscious *and* physical ability to interject our force, and acts of will; but no ability to eternally sustain or destroy anything. Our interjected forces of aligned and contrary polarity, to some small extent, acts to enforce or cancel out the purpose and appearances originally perceived by GOD and others. Yes, if our force is not contrary, it affirms GOD'S and others work. GOD must design and create all our required external perfection, and we must align in our proper sustaining force. So rather than hurt his creation, we all sustain it. This pleases and eases GOD'S work. Now using this body, souls can cut down trees and build shelters, and play with the largest quantum structures. And by self-rule, mankind has disrupted the perfect orders of life. Yet, the deep smaller faster quantum entanglement, forming the functioning aether, and the Holy of Holies, remain as created; these, no lesser soul has or can alter. Our projected patterned forces are absorbed into the existing created reality; but this is not anything we physically do, this is a godlike forceful ability apportioned to us, out of and from FATHER. GOD'S and mankind's projected affects upon the physical aspects of creation, are of immensely differing strengths, being at opposite ends of an immense scale, yet they are one quality of common force. Again, Entanglement, the Observer's Affect, the Quantum Zeno Affect, and many others, all point to the Conscious Creative Force Principle; all being quantum affects which do not fit a materialist understanding. In FATHER'S model of reality, we see creation and destruction are affected by our forceful entangled beliefs, through our relativity and our observer's effect. Entanglement is and has a far greater purposed understanding than materialist believe. We see mankind affects the quantum and physical world around us by projecting certainty or uncertainty; yet minus FATHER'S physical provision and allowance, we can do nothing. In this ever-escalating battle for souls, FATHER will soon allow powerful Holy Angels under CHRIST, and powerful Fallen Angels, to join his battle for this world itself, projecting great forceful intentions to alter the appearances of our realities. We see a temporary allowance for all souls to fight over, and alter FATHER'S creation, which serves his loving eternal purpose. The future horrors that soon come upon the earth, will cause most individuals to ask, how can GOD allow such cruelty and insanity to come upon the world. But this is exactly what individuals have chosen to do, deep in their souls. They think FATHER is cruel because he will not let them take his world, and reform it in their own desires. How insane is that? The coming Tribulation will be the loudest voice from GOD to date, saying he is there, and mankind can never govern or sustain themselves. We must see it now spiritually impossible for GOD to be heard by most souls, being so far apart in belief. I tell you an immutable truth; GOD is in no way compromising. GOD has no quality for wanting or accepting contrary imperfect forces; GOD is omniscient. GOD loves us in our returned measure, or they do not love us anymore. If we return to GOD more and more, GOD returns to us more and more. But as souls move further from GOD, they must love those same souls, less and less. Many lost souls moving away from GOD, can still reverse; but if one turns and receives HOLYSPIRIT'S understanding of GOD, and turns again to not love and want GOD'S ways; they are sealed as lost forever. In FATHER protecting his children from evil, he sees every evil soul, just like he sees Satan. We know FATHER'S plan brought and ordered all the unique size entities of conscious force below him. And we are all, even collectively, forever minute and subject to FATHER'S great size of forcefulness. The rebellious are odd, in how they tempt each other on in foolish pride towards their destruction. Does the size, reach, and forceful-measure of each soul determine the damage they can project in each moment? Yes. Mankind is only forceful enough to interpret and affect small local areas of reality through their desires. And of how much more, are the far more powerful angels capable. Yet if we love, submit, and connect to FATHER, he can project his will through us in mighty ways. As found in our example of CHRIST. John 14:10, "*BELIEVEST THOU NOT THAT I AM IN THE* ***FATHER****, AND THE FATHER IS IN ME? THE WORDS THAT I SPEAK UNTO YOU I SPEAK NOT OF MYSELF: BUT THE* ***FATHER*** *THAT DWELLETH IN ME, HE DOETH THE WORKS.*" John 10:34-38, "***JESUS*** *ANSWERED THEM, IS IT NOT WRITTEN IN YOUR LAW, I SAID, YE ARE* ***GODS****? IF HE CALLED THEM* ***GODS****, UNTO WHOM THE WORD OF* ***GOD*** *CAME, AND THE SCRIPTURE CANNOT BE*

BROKEN; SAY YE OF HIM, WHOM THE ***FATHER*** *HATH SANCTIFIED, AND SENT INTO THE WORLD, THOU BLASPHEMEST; BECAUSE I SAID, I AM THE* ***SON*** *OF* ***GOD****? IF I DO NOT THE WORKS OF MY* ***FATHER****, BELIEVE ME NOT. BUT IF I DO, THOUGH YE BELIEVE NOT ME, BELIEVE THE WORKS: THAT YE MAY KNOW, AND BELIEVE, THAT THE* ***FATHER*** *IS IN ME, AND I IN HIM.*" We see the desire of FATHER and HOLYSPIRIT worked *through* CHRIST as he existed in the flesh of man. Again, Psalms 82:6-7, "*THEY KNOW NOT, NEITHER WILL THEY UNDERSTAND; THEY WALK ON IN DARKNESS: ALL THE FOUNDATIONS OF THE EARTH ARE OUT OF COURSE. I HAVE SAID, YE ARE GODS; AND ALL OF YOU ARE CHILDREN OF THE MOST HIGH. BUT YE SHALL DIE LIKE MEN, AND FALL LIKE ONE OF THE PRINCES.*" Through CHRIST, in his suffering years on earth, FATHER, HOLYSPIRIT, and the Holy Angels restored the observed certainty of many things, to show that GOD exists. By these forces, CHRIST healed whomever he would, went wherever he wished, and even over-powered a storm. FATHER, at the beginning of this entire creation, did force, speak, and move, a portion of the physical points of his body, to become a grid-like sphere outside his self-perceived physical person, to be his aether. An aether still of him, that obeys his will, as he holds his creations between his finger-tips. By FATHER'S hands pulling back from sin, that spherical aether, and all creations within it are stretched, to soon be dissolved. We need only look around to see this is true by all that has occurred, and what is occurring, and what will occur by the sure words of Prophecy. In knowing our loving eternal GOD, we see an immense omniscient measure of what we are meant to know and be in our minute way. FATHER was once utterly alone; so, we see his reason, his loving motive, and the eternal purpose in FATHER to seek a shared *eternal* love between himself and other free-will souls. Do you see all the vast planning and effort, do you see the free-will probabilities, and the painful cost of choices that are now set against GOD, in so many lesser souls. Yet a loving FATHER, SON, and HOLYSPIRIT, endure their painful work, hoping and seeing beyond an uncertain extent of loss that will come by free-will. There was an all but non-existent chance, that all gifted souls would eternally love GOD, or that all would rebel. Has FATHER'S plan to find eternal love, been worth the eternal cost, over eternal loneliness. At minimum, FATHER'S first 13 of 665 gifted souls, chose eternal love. And all of the first 665 likely chose eternal love. But as is the nature of love, FATHER'S hope is for *as much* love as possible; so, he continues on towards the conclusion of finding that very last possible eternal love, through an ever-growing cost of ever-diminishing returns. The conscious and physical sacrifice and cost to FATHER to find that last tiniest eternal love, may in its contrary ways, equal and balance that cost of finding his first and greatest love. In this, creating a perfect circle and union of the beginning with the end, and the first with the last. So, we see a principle, being every eternal love, comes at a cost to FATHER. How valuable is *eternal* love, that GOD together pays so much for that smallest last eternal measure. I believe any loving omniscient soul, being in the same lonely situation as was FATHER, would choose the exact same course of action. We painfully see the endless confused anguish that will fill those who refuse to love GOD. We painfully see that as all reality unfolds, fewer and fewer of the lesser and lesser souls will choose to love GOD. The painful loss of love is felt and known by every soul on both sides. But the *rewards* of true eternal love, are only with

GOD. The many by loving and choosing a godless reality, become purely physical, and self-identify as purely physical creatures. These believe they are the body they sense, and falsely believe their awareness resides somewhere in the wet fat grey meat within the head. They're indoctrinated, as their desires and beliefs for this world grow firmer day by day. How can they hear, how can we tell them, how can they suddenly accept something so contrary to everything they now believe and desire? How can we watch them forever descend deeper into madness and delusion? These, whom the eternal poets might call "*the Fleurs Du Mal*". FATHER knew there *would* be pain and loss, in his search for eternal love. But FATHER did not and could not know to what *free-will* extent, that costs might be. Now FATHER, and we too, are seeing the unfolding extent of these cost. We see the choice FATHER made between searching for eternal love, or remaining in eternal lonely anguish; and we each must make this exact same choice. Do you choose eternal love, or fall in eternal lonely anguish? Do you truly sense the eternal side of the equation? Is this great eternal faithful love found in the fewer, worth the existence of the hate and divisiveness that has arisen in the many? Is chosen eternal loving life for some, worth the chosen eternal individual hell and death of the rest? Again, this equation was set in free-will choice, and beyond FATHER'S knowing control. Is the existence of GOD'S defined good, worth the cost of all free-willed denials of the same? Who is to blame, is it FATHER and his perfect plan, and great gifts? Or is it individual rejection? Who created the choice? But who made their choice? We know FATHER had to, and did gift and create our choice by the fact we are all here, and self-aware of our own desires and choices. We see others and ourselves having desires as we each choose our love or hate towards GOD. We see a complex conscious labor and process in each soul's free-will choices. A great love for GOD, *or* the vast rejection of it, are both staggering in their manifest measures. We see in every way, how the willful desirous force of GOD is immutable. One of those ways being that all CHRIST saves, by their own choice, become perfect and eternal as he is; and GOD loves them dearly. Now again, when we dream, we can fly, or fall, or walk on the ground. So, do we interpret what we see; or see what we interpret? The body's created entanglement to the earth mankind calls gravity, came at the initial designed creation of this world of life. These bodies are forcefully entangled with all reality by GOD, but not our soul. So, we bring this all back around to our conscious condition, as we need to be eternally entangled with GOD in belief and life; or be set adrift on our own untenable individual path. Most choose to abuse the body, and mold the world around them to fit their rebellious physical desires. We must understand that free-will means FATHER gifted us to make every single choice for ourself, and we are accountable. All evil choices will be possible, up until FATHER removes all choices for evil. We can be as the Holy Angels, or as Satan's Children who are demonic and antichrist; and no benign middle ground exists. CHRIST said we are for, or against him; all in, or all out. Our current desired chosen Godly or rebellious understanding, has been chosen by us bit by bit. Learned and believed as our every experience was interpreted through our core desires. Our every experience is perceived through our core desires to follow, or rebel against GOD. Why do *we* see how everything has come from GOD, but others see it has all come of itself? Why do we see an eternal loving purpose to it all, and others see no eternal purpose whatsoever? These are our choices made in seeing all things through our chosen desires. All souls who say they do not believe in GOD, if honest, must say they also have no desire for GOD. I have heard some dishonestly say, they wish GOD were there, but see no evidence; this is a deceitful soul. Could or would FATHER desire for us to remain weak against sin and unlearned of him? That also makes zero sense. FATHER through his loving desire, has created a process of choice. Which through free-will, we can grow to be one within the obedient mind of GOD. The eternal truth is there, and FATHER, JESUS and HOLYSPIRIT, obey this eternal truth faithfully and responsibly, as do all born of GOD. It is no more complicated than this. By the true interpretation of Scripture, HOLYSPIRIT shows us the truth of these two opposing paths of love, belief, and choice. And that the term reprobate, means a soul who by their desires, choices, and formed beliefs, has become so far and firmly lost in their confusion about reality and GOD, that they have made themselves unreachable. Or do *you* believe that all souls are redeemable up to, and beyond the very end of grace? Do you believe every soul will be redeemed, even Satan? Such beliefs are profound contradictions, as such souls will desire and always turn to destroy. Do you see that if you in any way contradict Scripture, you contradict GOD'S love? Satan's work is in denying GOD'S ways of eternal life, and teaching false doctrines that allow the perverse lifestyles found in Sodom and Gomorrah. Scripture says that all who contradict GOD, are children of the Devil. Please see it is blasphemy to choose and accept only certain verses, and self-interpret Scripture outside its context. GOD chose and watched over certain men to reiterate every word he once gave them; but not those who have now reinterpreted it for themselves. Do you hold any individual Biblical interpretations, making yourself the arbiter of who GOD is, what is GOD'S truth, and what is good and evil? This is the exact description of Satan's chosen ways. Lucifer, then one third of the Angels, then Eve, then Adam, and so on, chose their own way. Does GOD now look down and think like mankind? No. Does GOD now agree with whatever we decide? No! GOD is the sole judge of what we each have become, compared to all righteous omniscient eternal loving desires. JESUS is the standard that defines and preserves the saved soul, from the unredeemable soul. If FATHER sees me on Judgment Day, after all his loving efforts towards me, and says that though I tried, I am still flawed in my love, and I am not enough. I will know this must be true by GOD'S loving righteous omniscience. I will also know GOD is only hurting and grieved by this fact. And in this, I must yield to GOD'S supreme understanding above my own. I will know my core desires still held evil in some divisive way. I would then descend into the insanity of my separation and my own devolving conscious state. And I too, would eventually lose all love and truth. I would become utterly alone and filled to overflowing with confused emptiness, minus every richness of existence that might have been. I know GOD will be hurt by this, yet they must honor and preserve the true eternal love they now share. From our very first perception, on to our never-ending perceptions, it is we alone who *choose* our desires and reside within them. It may sound circular, but at our core, our love equals all our desires. If our love for GOD is true, we can only ascend into the eternal protection and provision of their Kingdom in a loving peace mankind cannot dream of, at the beginning of the eternal Kingdom of GOD. Let us in a true way, see this mass of neural wiring and connections of energetic firing points called the central and extended neural system with its vast web of tentacles throughout the body. If you believe in your eternal soul, you cannot also believe in any conscious attributes the godless sciences say emerge from this wet grey meat. The sciences to this very day, have not found the first trace or place of created perceptions or awareness in a brain. They can't describe what our perceptions or awareness even are. They have looked everywhere in every possible way, finding nothing more than mappable neural correlations; yet they hope and believe there is no soul. These simply preach a godless religion, founded upon false faith and hope. Asking us, like all worldly preachers, to have faith in *their* individual interpretation of reality. Why do many say they believe in GOD and our soul, yet believe a contradiction that their mind is in a physical brain? This brings confusion and division to any belief in GOD. Choose one now, are you self-aware by the wet meat of a brain, or by your eternal soul? It must be one, and it cannot be both. The brain is part of the body interface that returns to the dust. The brain is not filled with who we are, and resurrected before GOD. The wet meat of a brain does not ascend or descend with us, or hold our final station. The conscious force of our understanding self-aware soul, with all our created memories are one; a fact we must firmly avow now! All physical evidence seen from a proper perspective, shows souls are the only possibility. The body *and* soul, are not hardwired together as one quality. Science sees countless air-gap, across space, wireless signal jumps. Trillions of across gap frequency points of energetic information are forcefully sent and received, to and from our soul, making neurons, nerves, and every micro-structure having resonance, even microtubules, part of a vast two-way antenna interface system called the *human* body. And there are surely layers and sources of forceful communications in the abstract quantum realm, nearer our soul that enable and affect our awareness. There are unknown unidentified levels of receiving and sending forceful energetic patterns of information our soul uses to experience and control the body's interactions with the vast creation of GOD. Within our soul we experience awareness of everything physical from pain to pleasure, upon all our desires and emotions. Materialist ask, how can a learned knowing soul, forcefully align physical complexity? And we ask, how can unknowing particles and atoms, align themselves into the inexplicable complexities and systems of creative conscious awareness. They have no clue, yet they have faith, hope, and belief, that it just does. Then *they* force their godless faith and hope upon *our* children as being the truth of reality. No theory on any form of a material-based consciousness has ever found one tiny point of proof to support it. There are endless text books, filled with their religious desires, hopes, conjecture, physical miracles, and abracadabra rub it and wish, and a conscious genie pops out of the wet fat grey neural lamp. Their impish emperor has zero clothes. I see a slick salesmen, trying to sell me a bridge from nowhere to everywhere. Their every proffer, is a pure supposition of their godless faith. They try to find endless variations of going down the same godless material path, searching for any slightly new perspective in their shared false physical hopes. They refuse to see the fact, that it is their underlying hopes and godless beliefs, that form their too small too slow confused perspective. Most of these souls can never accept defeat, or stop searching for physical answers, after investing years of their life and reputations. The *politics* of science controls their religion; and a physical way of life has become their master. Many think the physical sciences are apolitical, but they above all other avenues in life, have always been on the front lines of pushing biased political beliefs. Long ago in times of wisdom, the politics of science were referred to as the dark arts. The name has changed, but not the darkness thereof. A student of political science, who says their individual beliefs do not affect their interpretations of data, is blind, or deceitful. They are just fallen souls, and have no higher uncommon mantle of steeled honor. Like most others, they seek a comfort of belief in life. They do not reside above, while the rest of us dwell below, in a realm of biased desired influences. Seeing their eternal soul, would trigger their fear of being wrong their whole life. A sociological fact, is it's easy to indoctrinate someone early on in life; but extremely difficult to later convince that soul they have been indoctrinated. Our first strongly held desires and beliefs can be deep and powerful. So, the more proud, self-centered, and invested a soul becomes in desired false beliefs, the more difficult any correction becomes. If you listen for it in their lectures, you hear materialist are trying more to deny the existence of the soul, than to show any form of physically manifested awareness. The political physical sciences try quite hard to avoid the truth. They keep looking deeper at quantum brain functions, yet they find nothing but energetic neural correlations, which only indicate physical "interface activity". Being no more than energetic patterned information, as it is formatted and exchanged both ways between two separate external differing and distinct qualities and quantities. For millennia they have searched the brain high and low, stem to stern, top to bottom, inside out, by every conceivable method, practice, and every strained logic and wish, and have found zero evidence of any physically formed awareness. It's a well-established scientific constant, and a most firm Principle, that no one can ever prove or find anything, unless it truly exists, or you look for it in the right place. You can search for a physically sourced awareness forever, that does not exist, in a place it could never be, and you will find. Materialists across the ages, spent their lives and great sums of money, relentlessly searching for their false material dream; and all died, not finding it. Their false abstract hope, exist *only* within the wave functions of their desired and built-up reveries, which seem real within them. But it is all non-existent beyond their abstract hopes and desires of their soul. It is all an unfirm dream, built under the distorting influence of their founding core desires *against* accepting the existence of GOD and the soul. Now, we can strongly feel a bodily connection, but unlike our soul, we know this body is not eternal, or ours. And *as* seen in the Quantum Zeno Effect, FATHER must systematically know, align, and sustain, *all* physical functions, or they falter. But if we add physical confusion to the body, FATHER does not increase his pre-set forceful patterned care for each physical platform's condition, so it degrades through its stages of deterioration or dying. All once perfect dynamic physical systems, perceived by FATHER, now residually function through stages of deterioration. The body's sensory systems are one-combined interface. Picture not the entire eye, but only all the sensory nerves of the eyes, and the nerves of the ears, nose, and taste buds and skin and muscles. Every sensing and responsive aspect in every cell of the body has the ways and means to receive and send forceful patterned information, and are part of a created interfacing system. See only this entire network of all sensing aspects of cells, microtubules, nerves, neurons, and more, and we see this is what our quantum soul knows and is adjoined to at all points. In not knowing this, the outer physical surface appearances will have far too great an impact on our perspective of belief. In all ways, our *desired* interpretations of our perceptions, form our reasons and purpose to exist. By our soul's desires and many experiences, we form our trust, which leads us to our beliefs as to the existence of GOD and reality. One cannot escape all their desired false beliefs, born of biased interpreted experiences without great eternal reasons. Free-will, as with all things is a two-edged sword. And sensing all life through a physical temporary perspective, one will strongly desire all that physically pleases, calms, satisfies, or thrills them to differing degrees. But one is then subject to, and controlled by these physical desires to the same degree. I tell you the truth, Heaven cannot exist eternal in the presence of desired contrary ways or beliefs. Free-will *does* exist, in that no soul can consciously force another to truly desire or believe anything against their own will. And free-will is not like a massive switch panel, where we can flip switches for our desires to on or off. We must know our strong, wrong or right desires and chosen beliefs, will push and pull us towards the same. And in lovingly longingly seeking GOD'S truth and understanding, false desires and beliefs can be corrected. FATHER gave us our free-will soul that from the start has learned and believed as it wished, be it towards self, or GOD; it is all just that simple. Early on, most all have confused desires, but we can learn, grow, and come to see the many destructions, pains, and losses, that come upon us and others by our wrongful desires. Through prayer and determination, past contrary desires and ways are rejected and denied, rather than accepted and lived. After admitting one's desires and ways are defined by GOD as contrary; to change requires a contrite determined effort, and avoidance of temptations. We all know if a crack addict wants to change, the first true step is to *flee* temptations by avoiding all addict friends and dealers. If one stays among sin and temptation, they are deceiving their self, and do not truly want to change. Yes, they may want to momentarily change on the surface, yet deep down, they still want to remain the same. Each individualized soul, in their own minute unknowing inadequacies, can see things as evil. But how can they know what is truly evil on their own, when every individual has differing definitions of evil. Most of mankind have blinded their selves into accepting most lesser to moderate evil as being benign. And the list of evil, now being redefined as acceptable and benign keeps growing. If enough people do it, it becomes acceptable; this is the way of a fallen world, unmoored from GOD. If a soul cannot see *why* something is evil, it does not make sense to them that it is evil. This is why we all must learn the omniscient eternal reasons for all of GOD'S ways; these good changes will then make perfect sense to us. We then find strength, seeing the eternal good our changes perpetuate. Most all souls hold reasons of self-interest, as to why murder is evil; and by this, mankind commonly accepts it is evil, while playing with its definition. But if *we* define anything as evil, by anything within ourselves, *we* are still playing a ruling god. If by an individual perspective, one believes a certain thing FATHER hates, is *not* evil, or they agree that a certain thing FATHER hates, *is* evil; these two beliefs are neither of GOD. We must know murder is evil for *one* reason only, because FATHER has said it as evil. This is what defines evil, not me, or anyone else. Now as the world population has exploded over several millennia, by this increase, souls hear of, and actually see more and more real and make-believe death and murder in this world. And this has caused souls to become more and more desensitized, accepting, and unmoved by death. A real desensitization to death is possible, just like all other things; and it brings an increasing loss of spirituality, and humanity. Never see GOD'S and your belief concerning evil, as being aligned in agreement, yet coming from two completely different souls, or sources. It is about becoming one with GOD, and agreeing *from* GOD'S eternal perspective. Only FATHER knows how to define the things of life that are both joyous and eternal, and only GOD who knows what destroys them. We put our every trust, fully and only in what GOD says; by this, we cannot and will not fail. In any other way or measure, all souls will fail. It is no paradox that souls ignorantly seek most, what they most want. If one *thinks* and *says* they want GOD more than this world, yet fails in trying to leave this world behind. Then their greatest want, turns out to be this world, and they did not realize what they truly wanted. If we see and understand that the life of JESUS defines all the good desires we can hold towards eternal life. Then for our best effort at over-coming all evil desires, we must learn and see why GOD has said those certain desires are destructive to eternal life. Our love for GOD is evidenced by our obedience, wrought in our understanding *why* evil is, as GOD defines. Why can't all souls see evil as it is defined by GOD? Do souls to differing extents, not care *enough* about what GOD believes? Many souls cannot get past the point of saying, why can't I just *agree* with GOD that murder is evil, but have my own reasons for why it is evil? Why must I know murder is evil for the same eternal reasons that GOD says it is evil? There are weak individual reasons that can be believed, that can never hold-up and stand for eternity, and there are eternal reasons that will stand. Can you grasp this difference? There are weaknesses that allow souls to form wrongful inner desires. Souls fall under a spell of their own desires, resisting less and less, and forming meandering justifications for their desires. I hope we all see it takes true loving responsibility, and years to correct all our wrongful desires with GOD'S guidance. And if we truly strive, and it is needed, GOD has promised us a thousand more years to build and polish our true love. Now we are eternal souls, so how many years this takes, is insignificant. No one in ignorance and lack of concern, can overcome their strong inner divisive desires forever, we must first believe GOD'S every word is true and right. To defeat the contrary desires we war against, we must study and understand them if we are to win. We must rise up in all the strength we have, to completely defeat the evil side of our self. We can only win if we rise up through eternal love, wisdom, and obedient emulation of CHRIST, remaking us into a much higher image and fashion of a child of GOD. There are many higher ways of love from GOD to cling to and draw from, that are centered in and around, not just the love from GOD, but from our circular eternal love from and *for* GOD. It is only our eternal loving responsible desire, born in intimately knowing GOD through *faithful* reciprocal sacrifice of self, (crucifying ourself daily) which cleanses and transforms all our deep wrongful desires. The *flesh* pulls us in a fierce desire towards the illusion that *it* is supreme. Let me give you the best definition of true individualism (it is a soul understanding reality for themselves). Yes, there are so many degrees of individualism, but a lesser measure of individualism, is not our goal, or GOD'S standard. Now consider how if you go to a theatre to experience a bright, complex, fast changing and thrilling plot in a movie, with exhilarating music playing upon all the keys of your physical desires. This brings you a temporary illusionary heightened sense, but when you leave, you cannot keep that sense forever, it fades and you are left with only echoes and a loss of that sense. Now seeing it is temporary, we see it enables an illusory process in which one will want and need more. So, we see only our Spiritual side enables our eternal reality, and the physical side of life is just a movie. If you try to walk out of this world's movie, what will you be to GOD? Will you still want and need more thrilling physical experiences? What desires do you want to feed, and how can they be fed and sustained eternally?

Choose the loving joyous peaceful satisfying *spiritual* reality which can and will be provided, over any individual physical illusion that will not be provided. We see the eternal truth, and not a timely fading fantasy, and *only* two directions for our conscious forceful awareness to travel. The one created straight narrow path, leads us towards an ever-expanding shared oneness in the loving ways of GOD, as all endless individual physically desired paths, run deeper and deeper into hell. These go in every individual direction *away* from GOD, towards an unseen ever-narrowing existence within a shell of one's self of being utterly alone. All the individual beliefs that souls label as being Christian, are destined for the same hell and death. All faithful loving responsibility towards eternal life is the only truth and force that can direct our soul. We see by every-single created purpose and reason in FATHER'S design, that all true love *in* his Children, draws and attracts them to this one shared belief and way. GOD said in Prov. 8:17, *"I LOVE THEM THAT LOVE ME; AND THOSE THAT SEEK ME EARLY SHALL FIND ME"*. Deut. 5:10, *"AND SHEWING MERCY UNTO THOUSANDS OF THEM THAT LOVE ME AND KEEP MY COMMANDMENTS*". GOD has clearly shown us why they cannot forever love any soul who will eternally disagree and deny their created eternal ways of life, and who instinctively need to persuade others to do the same. Worldly Churches bow to this world's evil ways more and more, as Satan fills more pulpits with his demons. Such false Preachers do not understand reality or Scripture, and they focus solely on the Bible's words of love, mercy, compassion, forgiveness, and hope. These false Preachers leave the what, how, why, when, and where to each individual's false religious interpretation. It has been said, and I agree, that evil false Churches, and many false religions and doctrines have been set up by Satan, to replace the true Church. We see the democratic consensus power of a lie, exists according to the number of souls who desire it, and believe it. But truth and wisdom do not exist by consensus; and they only exist in the fewer who intimately know and live by GOD'S created ways of eternal life. Now in our effort to live the truth of GOD, we are never forced or coerced. Yes, many try hard and fail, but all who do not try, fail. Lucifer was the first soul who in pure prideful envy, could not love GOD enough to live under them, where he was created and gifted to be. We see that there is nothing in, or of GOD, that is not part of, or in concert with bringing eternal love and life among many. All weak temporary loves, and all hates that war against every truth and aspect of GOD, are born in turning away, losing, and forgetting, the created perfect ways of GOD. It is a truth, that if we cannot agree with another soul completely, we cannot live with that soul in all eternal perfection. In perfect spherical reasoning, we only desire that which we see and understand; and we cannot understand anything we do not desire. It is impossible to completely know someone, unless both are in complete agreement. If we disagree with GOD in any single *eternal* aspect, we cannot intimately know, understand, or make every sense of GOD in that same certain *eternal* aspect. So, a division of understanding and belief, and a disagreement has appeared. If JESUS says he doesn't know us, or never knew us; this means we are not, and have never been *one* with him in his belief and ways. We are then individuals who can make no true eternal sense of, or to CHRIST. Matt 7:21-27, *"NOT EVERY ONE THAT SAITH UNTO ME, LORD, LORD, SHALL ENTER INTO THE KINGDOM OF HEAVEN; BUT HE THAT DOETH THE WILL OF MY FATHER WHICH IS IN HEAVEN. MANY WILL SAY TO ME IN THAT DAY, LORD, LORD, HAVE WE NOT*

PROPHESIED IN THY NAME? AND IN THY NAME HAVE CAST OUT DEVILS? AND IN THY NAME DONE MANY WONDERFUL WORKS? AND THEN WILL I PROFESS UNTO THEM, I NEVER KNEW YOU: DEPART FROM ME, YE THAT WORK INIQUITY. THEREFORE WHOSOEVER HEARETH THESE SAYINGS OF MINE, AND DOETH THEM, I WILL LIKEN HIM UNTO A WISE MAN, WHICH BUILT HIS HOUSE UPON A ROCK: AND THE RAIN DESCENDED, AND THE FLOODS CAME, AND THE WINDS BLEW, AND BEAT UPON THAT HOUSE; AND IT FELL NOT: FOR IT WAS FOUNDED UPON A ROCK. AND EVERY ONE THAT HEARETH THESE SAYINGS OF MINE, AND DOETH THEM NOT, SHALL BE LIKENED UNTO A FOOLISH MAN, WHICH BUILT HIS HOUSE UPON THE SAND: AND THE RAIN DESCENDED, AND THE FLOODS CAME, AND THE WINDS BLEW, AND BEAT UPON THAT HOUSE; AND IT FELL: AND GREAT WAS THE FALL OF IT." If CHRIST says he *never* knew someone, that means they were *never* forgiven; not once. In the above verses, JESUS talks about many who ask for forgiveness, and in their individual belief, thought they had been forgiven. Many believe that FATHER *must* forgive whatever they ask in JESUS name, like it is a real magic incantation. But we read that many were not forgiven, yet they believed they were. All such souls, actually take the LORD'S name in vain. Scripture says, GOD must see a true contrite repentance of our evil as a pre-requisite in our asking for forgiveness. Repent means to turn from, so repentance of all evil, means to turn away from all evil; not some, but all. And again, GOD alone defines evil. Until we hear CHRIST say enter, "thou good and faithful servant", we wait in our loving responsible faith, and his promise, and our growing obedience to his every truth. Amen. A lesser known and understood aspect of eternal spiritual love, is that its bright glorious facets are nothing like the dull pitted fading facets most hold as a worldly, or ethos love. Being a fallen lesser imperfect impure fading and more physical side of love. The world instinctively broadens and dilutes all great meaningful words. And the words *eternal love* to GOD, means by our worship, we responsibly put the eternal needs and welfare of GOD'S Family and Kingdom before self without fail. Our measure and strength of eternal love is found in how far we put GOD'S Family and Kingdom above our worldly family, friends, ourself, and possessions. Our eternal Godly spiritual facets, are found in our selfless sacrifice towards, not this world, but every word and aspect of GOD'S Kingdom. Only within and for GOD'S Kingdom, can there arise our eternal responsible faithfulness to grow to become one in the eternal perfection of GOD. These truths are all highly illuminated in the loving words of GOD; and we see the brightest facets of CHRIST, are expressed towards FATHER. Through these last days, and the coming Millennial Reign, these brilliant facets can be cut and polished within ourselves. In this way, this highest love we receive, can be returned back to GOD in our fullness of truly being one with them in belief. How do souls find an ability to not love the source, the beginning, and creator of love and life itself? How does one not love most, the eternal source of life and love? How can or do any, *not* love GOD the most, who created every founding purpose and reason and way for love to exist? Which commandments did Eve and then Adam violate in eating that piece of fruit? First, they put their desires for self above GOD, breaking the first and greatest commandment. They were also unfaithful to GOD, being spiritual adultery, so in that, they broke commandment seven. They stole something not given, so, they broke commandment eight. All sin leads to death, so they broke commandment six. We can go on to show that they broke all of GOD'S commandments in one rebellious act. You may ask why we choose in this book to describe our every choice as a conscious formulation? Well, a formula is simply a list of the required ingredients, and a specific order *and* measured way, those ingredients need to be introduced, for a specified outcome; and the term formulation, best defines our every choice. All souls sealed in GOD'S perfect love, direct their free-will choices for the good of GOD'S Kingdom. Let us underscore the fact that the Trinity of GOD have never done a single evil, and GOD has never hated another soul first, or any good thing. GOD has gifted and *created* the free-will potential for love, because it requires allowing its free-willed rejection. FATHER gifted, created, and has allowed a choice potential for every free-will form of love, hate, good, and evil between all souls. No soul gifted by GOD, can alter GOD'S defined attributes in all eternal things; none have this power. But all have the free-will to form their own believed divisive definitions, and to deceive themselves and others. For the Children of GOD, the highest joys of love are no mystery. From our first understanding of an eternal commitment to love, GOD'S arms are open, their hands extended, they reciprocate every possible feeling we extend. GOD recognizes our eternal desires when we are lost in confusion and illusion, and GOD only wants us to come out of it, and back into their light. The great pains associated with love are no mystery, for in our efforts towards love, the pains of free-willed rejection and tragic sudden loss of love are experienced and endured by all. Again, by seeing the whole of all creation as one spiritually entangled purpose under FATHER, we see the eternal and the perfect become one concept. Every imperfect choice is antithetical to all that is eternal, and all imperfections are aspects of individual hell and death. Such desires form every chosen imperfection, and turn GOD'S love away. Most souls never experience any true eternal spiritual life. This world is *nothing* like it was first perceived to be, yet a perfect spiritual counter balance is provided by GOD in all things if we seek it. I tell you truly, a Preacher, Rabbi, Priest, or Teacher's works, are not judged as good by the thoughts and desires within the hearts of souls who come near to hear. It is in those souls *new,* never before known, Holy understanding and conviction in their hearts, when they then go on their way. By this, we see if a Preacher, Rabbi, Priest, or Teacher, truly serve GOD. It is never what we physically see, hear, or feel, in this world, but only the desired reasoned spiritual choices we and others make for the ways of this world, or the next. Now we ask all souls, do you see debilitating physical or neurological abnormalities and deficiencies as a curse, or a blessing. Well, nothing physical that occurs, is ever in itself a curse or a blessing. It is always a blessing if a soul looks and moves past that occurrence, and spiritually moves on to GOD in love and hope. And it is always a curse, if a soul dwells in and on the occurrence, and allow it to drag them down into hell. This is the only way to view blessings and curses in the flow of reality. In this we see all spiritual blessings and curses, are formed within us, between ourselves and GOD. Most think a person winning a lottery, received a blessing; but in the fullness of that money's effects on their soul, it may be the curse that seals them as a reprobate. Every conscious entity has their force of will, yet it is gifted to us all with a limited physical expression for very good reasons. Look at the current growing level of pain and evil in this world. If GOD gave us *more* physical ability, that pain and physical destruction would increase accordingly. We know these three things: one, the body is not our own and the life within it comes not from us; two, our eternal soul requires a complex physical interfacing platform/body to richly sense a certain level of the information radiating from FATHER'S external physical perceptions, (the universe and the fullness thereof); three, these connections for our soul's awareness, can be disconnected and reconnected, or then connected to another suitable sensing body, in as many times and places as GOD'S perceptive and creative control of each body, causes it to occur. And by these three *created* aspects, we define Soul Dualism. Science has invented a spectrum of definitions to include false forms of dualism to blur the fact that the conscious soul plays such a forceful evident role in creating our perceptions of reality. Why? Well, perceptions are literally our only evidence of reality. And Materialist cannot deny what they call the hard problem, or the mind-body problem. They desperately want and need to show life and our awareness as one closed purely physical system, but they can't, because it is not. They say love and sorrow are not real, that they are meaningless electro-chemical reactions towards propagation and self-preservation. But tell me *why* and *how,* would any purely physical thing, governed by entropy and chance, have an inexplicable driving force of physical self-awareness and a goal of self-preservation and propagation. Look at a physical living tree, does it *know* and *desire* to perpetuate its existence. Look at anything purely physical and ask yourself, how can it know it exists, and how can it desire to continue? Does animation bring awareness, or does awareness bring animation? Please see that all knowing and desiring to exist, is only, and eternally a matter of our soul's awareness. Materialist believe that in the future, love can be physically made in a laboratory and put into a pill. But to believe that, they must also believe that self-awareness and desire are no more than electrochemical information passing through a wet piece of fatty grey meat. They demand we do not doubt their faith filled hopes and wishes. And if we do, they call us science deniers, when they have no real science. They demand we allow them to teach and form and control what we and our children believe. They speak from a devious extremely exaggerated temporary physical perspectives, yet offer zero conscious specifics or proofs. But they know the masses, desire to hear and accept their lies. Show me any step-by-step process that can create perceptions and our awareness in brain meat. They come at every deep question about awareness, from many deceptive angles of supposition and personal belief. They preach an unprofitable faith; and if cornered, they do admit they truly have not the slightest idea what perceptions are, or our self-awareness; or where these are and how we create our perceptions, or memorialize them, or recall and reperceive memorialized perceptions. From our first awareness, each soul comes into existence at hearing a whisper saying "I am GOD" having only their newly gifted inherent ability to desire and forcefully perceive, create, indwell, interpret, and then choose to define and build our awareness of belief for our own internal believed existence. We, through our created perceptions and desired interpretation of them, begin to form our desired chosen understanding, and embraced belief of reality. Mankind's chosen perspectives of understanding and belief, are products of their desired interpretations of the sum of their perceived experiences. So, truth and knowledge for most, are an individual subjective truth and knowledge. At the forefront and surface of each individual's accumulated past subjective experiences, arises their current interpreted belief system. Yet high above all individual beliefs of understanding, we see something very different, called the omniscient wisdom and belief of GOD. In the beginning, the term "wisdom" only and always meant GOD'S original purposed understanding of reality, which declares its perfect design and designer. This wisdom once stood alone as the highest order of understanding. But now we see for most all definitive words and terms, including wisdom, they are being constantly redefined, as each generation loves and seeks to change and control the meaning of life. This is a generational insanity that pushes and pulls at confusion and divisiveness. It is part of the rebellious nature of the younger more foolish, as they wrestle the world away from the older less foolish. By this, each generation falls and fails more and more, winding up further from GOD in all their pursuits of belief and understandings. Mankind has been surrendering all understanding and the meaning of life itself. En-masse, the young believe themselves wiser than those before them; they do not see their new *so-called* wisdom, is no more than rebellious desires. If they desire it, to them it must be true. The first foundation of rebellion, is to believe the young know more than the old. If one always consciously faces the wrong physical way, they must always see, learn, accept, and believe untrue things that do not come from GOD. There is GOD'S true knowledge, and endless untrue knowledge.

There is GOD'S one true understanding, and endless untrue understandings. And there is only one true wisdom that comes from HOLYSPIRIT. Now all doctrines of societal victimization and equity are purely physical, and non-spiritual; yet the fewer souls still accept their accountability for creating most of their worldly problems. Each new fallen generation creates their own new rebellious unstable changing emotional and illogical systems and rules for playing out their fast-failing existence. In *all* countries and places of this world, their new desires and ways migrate away from GOD, personal responsibility, and liberty, towards liberal-minded zombies, who will one day soon, see and choose to put the prophesied Antichrist in power. Now the United States was never a truly Holy place, but on a spiritual level, it was closer and far more just when first founded for and by a people under Judeo-Christian laws and province. Founding patriot, John Adams once said, "Our Constitution was made for a moral and religious people, it is wholly inadequate to the government of any other." And Benjamin Franklin said, "Only a virtuous people are capable of freedom. As nations become corrupt and vicious, they have more need of masters." There are many similar quotes from many of our founders. This country's founders did not consider it mankind's place to control its Judeo-Christian people, or to put much power in a central government. The Constitution is a good conscious formula for a moral people who live under GOD. But the Liberal Marxist Demo(n)crats do everything possible to destroy our moral Judeo-Christian ways and understanding; and take away our highest national liberties and protections. This liberal conscious formula is an evil that brings only destruction. Since the founding of the U.S., all Godly moral souls have fought to preserve and give freedom and faithful responsibility to all; but liberals seek a broad godless interpretation. The Demo(n)crats adopted the godless racism of Darwin and Eugenics, to control the souls, Republicans freed from slavery. This country has gone in two directions: one under each soul's good religious moral patriotic traits, to live under GOD to overcome all evil within their nature; the other is secular and liberal, being set loose morally and emotionally, denying the greater good of GOD exist. Both sides now live in distrust of the other, being divided on almost everything, which brings a vicious institutional corruption in liberals that cannot be stopped or reversed from without. This is the prophesied last days rising global organism of the Spirit of Antichrist, which is now vast and complex, and it can't be reversed without the people turning back to GOD. Liberalism exists in individualism, and individualism is born in one's denial of strong Biblical standards being taught or encouraged. Liberal policies demoralize souls; meaning such policies destroy a soul's morals. And in demonizing and demoralizing a country, this destroys national strength, hope, and promise within the young, leaving them nowhere to turn but to liberal Marxist globalism. By these efforts, a country is divided and made weak and unaware of all rising external and internal threats. Globalist control, now includes every form of left leaning media. The definition of being woke, is what spiritually and emotionally happens to people, when the most *in*sane grab power, and turn to censor and cancel the life of the less insane. Then, one by one, these liberals destroy each other until the one most insane liberal soul is the only one left. Being a soul so confused, that they hate everything and everyone, and disagree with every true reason and purpose to exist. In a strange fractured unison, the woke relentlessly tell any and every lie over and over. The woke always cheat, and take anything they can. The woke seek to destroy every normalcy to increase their power. The woke are the leading edge of an insane perspective. The eternal, and temporary perspectives, are contrary views of trying to understand reality. A temporary perspective sees reality as existing between physical birth and the grave. Their temporary beliefs, purposes, reasons, and compassions, reside within this small slow temporary physical frame. When such souls ponder if there may be something beyond the grave, they imagine a purely physical fantasy, that centers on their individual desires, lacking any common spiritual purpose or reason. Ecclesiastes 3:9-11, *"WHAT PROFIT HATH HE THAT WORKETH IN THAT WHEREIN HE LABOURETH? I HAVE SEEN THE TRAVAIL, WHICH GOD HATH GIVEN TO THE SONS OF MEN TO BE EXERCISED IN IT. HE HATH MADE EVERY THING BEAUTIFUL IN HIS TIME: ALSO HE HATH SET THE WORLD IN THEIR HEART, SO THAT NO MAN CAN FIND OUT THE WORK THAT GOD MAKETH FROM THE BEGINNING TO THE END."* The eternal perspective is not of mankind, it is centered upon all the entangled eternal familial aspects of our forceful purpose that flows through all words, love, life, and light of GOD, sustaining a created shared eternal reality. Globalist liberals, control most all Bureaucratic and Autocratic tentacles of government, including the education system. They control and censor most all forms of Media. They control most all Corporate and Medical regulations. They fill the hierarchy of the FBI and all Intelligence Agencies. They are filling the Command Structures of our Military, Police, and Pulpits. But until they fill the rank and file of all these institutions, and confiscate all firearms, they must work in a subtle way to conceal their ultimate goal, even from their own blind followers. They must keep hatred stirred up against those trying to save this country and its souls. Short of tremendous death and destruction, this level of structured evil can't be stopped, it can be slowed and pushed back for a while, with the right person as President. A world leader is a helm and a rudder, who must face strong contrary currents, storms, and fierce headwinds. The hope for any country with freedom of religious exercise, lies in its leaders being souls who can stand strong, while being lied about and demonized by the true demons. We need a soul of great strength and the will to use it, a soul of great courage and conviction who knows and loves every founding principle of the Judeo-Christian way, having great support from the people. GOD send us, and conserve such rare warriors. Open people's eyes to stand with them, granting us some years of rest. Amen. We know we face the considerations of chaos, disorder, and chance, mingled in against our free-will ways of life, and we know we face many temptations to be avoided. Think of the Garden of GOD, were Adam and Eve could walk by and see the forbidden tree, yet it could also be easily avoided. Scripture says it was set in the midst, or middle, so it was avoidable. Did that *center* point of the Garden, represent the weakness and confusion of wanting it all? Did that point represent the fact, we can't have everything? Do our great battles dwell at our center in wanting it all? In this fallen conscious and physical existence, evil conscious and physical aspects must be expected, avoided, and overcome. We see chance brings beneficial and hurtful *physical* outcomes to differing degrees here and there, causing many to see and claim a false excuse for their spiritual failures. Chance has its allures in giving many a false hope of gambling to circumvent the rules. Both GOD'S external creation, and the allowance of all conscious free-will, are both necessary to allow all unforeseen chosen loving and hateful conscious elements to exist. The desired projected force of our eternal love, cannot be *fore*seen by the desired force of GOD'S created perception, or it will have been his will that created it. But it can be seen, known, and felt, afterwards. Real true love must be longed for and freely given; it must be known and projected by each soul on their own, prior to the others knowledge. Only then can GOD and we know our freely held love towards others and their love towards us. GOD cannot make anyone love him, any more than you or I could do the same. This is the eternal law of free-will love. True love is what GOD desires and seeks, and only in this way are we not alone. This is in all GOD'S words of forewarning, but evil comes quickly before GOD'S words can root and set, trying to blur the lines of what is truly meant. Evil tries to change all that is true and right, into what is false and wrong, as evil conscious formulas steer all those *willing* or *wanting* to be steered. It is not a hard thing to do, they just exaggerate, lie, and add emphasis and dire to certain concerns, making them more prominent than others. This divisive practice, breeds anger in weak followers against any and all opposition. Now the Biblical understanding of the word "lust" means to strongly want anything in this world for selfish reasons or pleasureb being in unison with the word "covet" meaning to seek to have what the eye beholds in lust. To covet, or lust, is to see and want anything above the fewer aspects required to sustain and propagate physical life, and serve GOD'S great spiritual purpose. The physical body was created to be an operating platform for learning and living a Godly spiritual life. And now liberals desire to take earned wealth from some, and give it to others, to buy their votes. Those who lose or gain wealth through elections, can forever change places; but an ever-growing injustice is engrained. Most souls caught up in their physical struggle, have lost all ability to communicate with GOD, and so, very little Godly good is ever truly done in this world. The evil leaders in this world manipulate and shuffle the deck of misery within it all. Vast powerful political alliances become great machines upon the earth that plow, and mow down the hopes of multitudes of unseen people, in their self-righteous zeal to remake GOD'S Creation in their own evil image. We see how this liberal insanity, taken to its final conclusion, will cancel and erase at least half of all language, understanding, history, art, and *all* Godly love and unity. Leaving souls filled with only animal zombie lust, and a their insanity. For an example of liberal zombie policing of words and their meanings, a beautiful Celine Dion song titled "The Power of Love" was recently attacked for its lyrics, that use the words Lady, and Man. For these same words, Christian Conservatives love this song's beautiful Biblical contextual meaning. We know social zombies couldn't care less that the majority of most country are offended constantly by the venomous stench oozing from their pores, and the hate being vomited from their lips. I cannot imagine the stores of fury building within GOD against all these souls. These in their every single instance and every breath, have defiled the world which GOD has made. These souls are trapped in the hell of their own temporary individual perspective, working to propagate among the masses, all that is abominable, dark, and a lie; and all in the sight of GOD. What name could be given to something so evil? FATHER calls it the Spirit of Antichrist. Every contrary word, every contrary act, thought, or belief, not aligned with the ways authored by GOD, are a part of the Spirit of Antichrist, it all being against CHRIST. What could be the coming final justice, passed upon all who alter and destroy GOD'S good great original work, ways, and purpose. This growing antichrist spirit, is preparing the way for the actual embodied Antichrist. But FATHER'S

Judgment will surely come upon all who take to themselves the work of molding the beliefs of the people against his ways. These lost souls, who by their desires and efforts, turn mankind's perception downward towards the great abyss of individual confused belief, GOD calls hell. These souls howl and rage at their every perceived injustice in the physical world, while their every effort and breath are spent trying to destroy and replace every good loving reason for GOD to sustain this reality in the first place. We need to see how our shared Godly responsibility is set in creation's orders, and how liberals destroy this order, causing responsibility to fail. Remember that in the Garden, GOD brought Adam first, and taught him alone for we don't know how long. Adam was gifted in strength, and was taught his purpose and place of responsibility. Eve was brought later in her lesser strength to help Adam, and was placed under his responsibility. Adam was not placed in any arbitrary ruling Godlike position over her, yet he was created over her in the order of responsibility towards GOD'S created ways. Eve was a *following* creation, in FATHER'S lesser orders of strength and responsibility. This was and is GOD'S good doing, so, let each man or woman reject their Godly responsibility and created place at their own peril. Please note; that if *you* move outside of FATHER'S ordered stewardship and his eternal perspective, all his created orders still remain. But they become unseen, and of no understood consequence to those blinded by a temporary perspective, and so, GOD'S truth is no longer understood or desired. At our foundation, a soul has only two basic ways to account for their existence. One is our certain projected forceful awareness and all uncertain physical aspects, are eternally bound together under conscious primacy; bringing all physical complexity we now see. Or two, is to believe that countless prior physical miracles, mysticisms, and an unknowing physical lawful design towards great complexity, just existed within a pure random physical nothing, or something, causing all firm physical certainties to form on their own; and then a far greater undefinable and inexplicable animated self-awareness, just miraculously appeared. Mankind barely sees 6% of the physical creation through their small slow perspectives and false beliefs. They see and admit, that at the edges of all they see, their laws and constants don't apply; just like when they look into the quantum. These are the upper and lower true boundaries of reality, fallen mankind cannot correctly see or understand from their chosen slow small temporary physical perspective and beliefs. All the things that define and divide reality, are truly conceptual and spiritual. Now no soul of mankind, can create the inner perceptions of another; but we can influence other's beliefs, and their environment. Souls can discuss, share, and work to influence the perceptions of others in many forceful ways. And if forceful enough, like the angels, and if allowed by FATHER, an angel can literally project external forceful patterned information into our environment, to be sensed by the body. By this, causing us to physically see exactly what that angel projects for us to see, thereby affecting our perception. In the Bible, GOD says this powerful affect will return to earth in the last days. An altered visual of what we physically experience can come to us in this way, but any change in our belief, do to such an experience, must come by changing our core desire. Two souls sharing a certain belief, can witness an event and hear certain words, and one soul may be diverted from that shared belief, but the other may hold steady; but both by desire alone. If I see fire come down out of the sky at the command of someone who says he is a messenger of GOD; but he speaks and acts contrary to what GOD commands. Is my belief altered by what I see? No. By my knowing desire I see a forceful lying spirit; and I will know *this* because Biblical Prophecy says a fallen angel will be allowed to do this very thing, and he will deceive many. It is a future blessing to study prophecy, and to know what comes. One who does not know what GOD has said will come, can be easily deceived. Eve, for a certain undisclosed number of years, knew the forbidden tree was evil and death, as GOD had said; so, she dared not touch or eat. For reasons of love and trust, we see Eve believed and was perfect, up until she lost something strong and old, or found something strong and new in her desires, and she fell from her perfection at around her six thousandth year of life. We read how her physical perception of that tree and fruit changed, but it was her core desire that changed first, after hearing and wanting the lies of Satan to be true. That fruit looked evil for so long, and then by her changed desires, it looked good. By her very first individual desired interpretation, she saw the fruit that always looked evil before, now looked good to eat; and she desired to believe Satan's words, that it would make her as wise as GOD. Through desire, temptation, and curiosity, her Godly desire weakened and changed, then her understanding of evil changed from what was taught her by GOD. Again, we see within *all* souls, only one form and essence of force, called the consciously projected force of free-will, being our force of desire. We see a most-enigmatic phenomena, which rests at the foundation of all creativity, and all created reality. We clearly see that through free-will, evil core desires can bridle any soul, pulling them down into their own endless ways of hell and death. There is inner self-deception, and all outward deceptions projected towards us all. The disguised personas of souls walking around, trying to appear as desirable in this world, or at least acceptable in this world. Wanting to be seen as something they are not meant to be by GOD. And within themselves, they consciously disguise others; in that they see others not as they are, but as they desire to see them. We have all tried to warn someone about another who was in their life, having no good intent; yet they refuse to hear us, or see it until too late. How can a random weak way of understanding life and reality, be a foundation of anything growing towards a perfect eternal life? It cannot, ever! In souls that hold no desire to live in GOD'S perfection, there resides a constant reordering of devolving wrongful desires. Each evil desire, in error, is thought to be good; yet each evil desire will eventually be boring, and no longer pleasing. Evil that is now desired, must diminish and be replaced by other endless evil desires. Remember that to live or exist, rests in our *creating* our perceptions of existence. And minus the physical half of our existence, no conscious forceful soul can form their *created* existence through desirous force alone. We must physically create our perceptions, or we cease to exist until we can, or do. We find the *true* reality through following GOD, or endless false realities in worshipping the substantive perceptions of this world we create. We control and chase our willful desires, and we create our perceptions *without* effort, by what can only be a newly generated and intended projected force. But what if souls do grow or shrink according to their choices; what if soul's falling, is actually synonymous with shrinking? This may sound like it could violate mankind's slow small laws of conservation; but even mankind says their slow small physical laws and theories do not apply to force. And we see they do not apply to our souls. Science sees force weaken, but cannot define what it is, or how it is diminished. They want to define force as a result of certain physical actions, being a caused affect, and not an essence itself; yet they have zero evidence or proof of this.

We know that force sets all their laws, and that force is not controlled or set by those laws. To whom or what do they, and we, ascribe to be the first supreme quality in forming reality? What are the choices? Do we literally see matter and energy? Do we literally see force? Do we literally see, hear, and touch reality? No to all of these. We only create our perceived awareness of reality; yet all these immutable facts, can't save many from their materialist delusions. Why and how can a soul want to be fooled so much, and be so blinded? It is a *most* enigmatic-phenomena indeed. *If* our *adequate* connections to sensed information, and the aether, should cease, how is our existence found, it too, has ceased. Not our soul itself, but our created awareness has ceased. Physical sleep and death are intermittent phases where we temporarily lose and lack our existence. We temporarily do not exist, as our awareness of our created perceptions, is our existence. There is no more to reality than this. So, sleep or death of the body, are different to our soul in only two ways. After nightly sleep we regain our rich connections to the *same* body many times, within the same external physical reality. Physical death is where we awake connected to a new body in a differing reality. We are all destined for an other-worldly eternal good path, or a never-ending evil path. Beyond doubt we all will awake connected to a new eternal body. Those finding GOD'S higher love will wake to his true loving joyous peaceful ways. But the great many, wake to their own created inner envious separation between all souls. These will eternally fall and be lost in ever-changing confusion, and never come to the anchoring truth of GOD. Their final enduring curse, is to become as pure an individual as is possible; finding the worst solitude of all possible existence, knowing only their own lies, existing as a pure lie. Hearing only lies they never understand, straining to speak only lies and never being heard or understood. Having lost all outward physical power and pleasurable senses of touch and oneness of shared belief and understanding with anyone; being an eternal generator of desire, trapped within endless unfulfilled wants and needs, who can never again sleep. We all will never again be temporarily disconnected from our awareness after we reach our eternal destination. The lost are trapped in the confused desired beliefs of their own created hell, set in a separate aether within the inescapable fusing conscious prison called the lake of fire. They will silently wail for death and sleep, but never experience such respite. This is *no*thing

FATHER does to, or within them; he only takes back what he created, and they abused, which was never theirs in the first place. Their misery is self-inflicted as the inevitable eternal result of all *individual* desired ways of life. From their temporary physical perspective, they did not care or see where it all led, in the eternal. They did not want to see how and why they will come to hate so much, the things they now desire; things that now taunt and haunt them without end. In this current temporary earthly realm, many souls choose endless hours of sleep to escape depression and despair; such escape is *im*possible in the after-life. Those who want to be a god over their own beliefs, will be exactly that, trapped in a cycle of wanting endless unattainable unmet desires. Their chosen core desires, in the end become a wrathful god over their individual reality, and not one thing more. They will have separated *themselves* in increasing misunderstanding between their selves, GOD, and everyone else. FATHER cannot delete, annihilate, or reclaim the essence of these lost souls that were eternal accountable gifts, being now contrary and defiled. Their own created mind of lonely pain must now eternally exist, or, the eternal value and cost of GOD'S eternal love will not exist, and reality becomes meaningless.

Lost souls build a confused solitude that is very different, and maybe worse than the loneliness FATHER once had to bear for eternity. Eternity is not measured going backwards and forwards in any linear fashion, it is endless in every spherical direction. Most souls desire to see reality in a linear way, even eternity, which is absurd. So, they can't truly see or understand anything properly. This is the fallen separation of individuals FATHER calls hell and death that he warns about throughout Scripture. Who can imagine descending forever deeper into a dark singular understanding of reality, not shared or known to anyone but themselves. Seeing other countless souls and unfeeling bodies, roiling in the currents of all their evil perceptions. Being souls who can only writhe in their utter individual confused disagreement and anger towards each other forever. All while the perfect formula of GOD creates and finds a family that shares their unifying anchoring eternal belief of loving joyous omniscience, being all who desire this way of secured oneness. To be one with GOD in life, we start with the wisdom that GOD has perceived all creations for a loving, perfect, eternal purpose. We see why and how this purpose requires we share one belief of life, for it to be eternal. An eternal unchanging belief, sounds unpleasant to many, yet knowing this final truth is only assuring to *us*. All things true and eternal originate from GOD, so we align with them. The truth is not just what we know or tell, the perfect truth is the sustainable created way and purpose of all the things GOD has made. The "one pure truth" is the definition of the original methods, designs, and purpose of all creation for GOD'S love. For any soul to live outside of the true and original loving created ways of GOD, is to live outside the truth that sustains all creation, and live a life that is a lie. We can eternally live within the pure truth, or forever die within our own endless lies. We can also for our moment in this world of choosing our beliefs, live in a multitude of religious mixtures, that are still outside the pure perfect truth. Using an exhaustive KJV Bible Concordance, please pray and study every Scripture containing the WORD "Truth". You will see that to be found in the truth, means to live as FATHER created us to live among all Principles of his Creation. This concordance study works for finding FATHER'S *meaning* for words like: perfect, love, faithful, obedient, sin, repentance, and others. We all at a point must ask what is omniscience. To define this Godly conscious attribute, it is best to start *not* at the physical beginning of creation as most now see it, but before *all*, that now exists, when FATHER existed all alone in his awareness and physicality of a single solitary personage. And so, we see a singular experiential loneliness, is part of FATHER'S omniscience in knowing all things. Meaning; not JESUS, or HOLYSPIRIT, but FATHER alone holds this loneliest understanding. Yes, eternal loneliness is our only contrast for defining and understanding what is eternal love. But the meaning of love in this temporary world is shattered into endless individual variation and distortions. Yet within GOD there is a perfect eternal desire for love, and a desire to serve only an *eternal* purpose. Perfect eternal love, can never be known or defined as having individual imperfection and inadequacy, which are the dying physical fading aspects of the rebellious. To even try to understand an eternal reality, or eternal love, or loneliness, you must know beyond doubt that you are an eternal soul, capable, or susceptible, to these two eternal states of mind. And that if we live and remain outside of GOD'S binding desires, we will be alone forever. By our knowing this, it raises our first longing fearful desire for eternal love. This knowing, turns all our desires towards all that is perfect and eternal. What if I witness a few things about someone new to me, that I see in a positive way; for example, maybe I see a young alert and talented singer. So, at first encounter, there is nothing negative in my knowing this person, so, all my feelings towards them are pleasant. In not knowing anything negative, the sentiments towards their great talent, may cause a room to applaud and cheer and have a feel of hope for their future? But learning more about them, we find a self-centered spoiled soul; so, why is that hope and desire towards them diminished? All desires, including love, grow or die in getting to know someone, altering our initial understanding. So, we must know that in FATHER being the perfect and eternal sustainer of reality, as he learns about us, he has no desire to find an imperfect *temporary* love that will be lost. The eternal existing perfection of GOD is the only perimeter, parameter, and foundation, for all things eternal, including eternal love. Nothing individual and imperfect, can sustain itself at all, let alone forever. Individuals project a need, and demand for sympathy, wanting to be understood and loved in the way they understand and love. But in the loving ways of the eternal, it is impossible for all individuals to love all other individuals. The only way to be eternally loved in a way we all know and understand, is for us all to be one in GOD'S belief of understanding. Our good life is brought through our good emotions *and* logic. Do you grasp that you are gifted and created in a such a measure, that you can truly know GOD and all reality. Yet secular science, is proud of insufficiently sensing, and falsely interpreting just 6% of the physical half of reality they slowly and minutely sense. Worldly understandings of love are temporary and impure, as worldly loves all fade. We see fewer and fewer loves that endure for fifty, sixty, seventy years or more; and most that do, are flawed, weak, and failing. If a strong worldly love lasted a thousand years, it will still fade to indifference and other phases, as each soul changes, until it eventually becomes hate. All souls can look upon those they at this time now love and say how could I ever come to hate them; yet this awaits all individuals lost in endless time and disorder, lacking true eternal mooring. Sadly, as mankind now believes the universe will slowly drift apart into less and less forever; all lost souls will in that way, drift apart in a growing confused hate forever. Hopefully you grasp the necessity of one shared understanding and eternal *anchor* of belief between us all. FATHER is the only eternal anchor to truth, love, and sanity, that exists. By GOD'S loving sustaining omniscience, being entwined through our every belief, we see each other in the original created ways of life, and love each other in one eternal way, and never desire the destruction of imperfection. Secular mankind's accepted understandings, reveal the interpretation they hold of reality, seen through their fear and hope that all things including their awareness, are *temporary* and physical, and that nothing is eternal. Most can't comprehend being aware for eternity; and yet, if not for nightly sleep, no one could begin to comprehend losing their eternal awareness. Looking at this world's nightmare, we see why one truth of life, can never be agreed upon among all the individual antithetical beliefs of mankind. We must know we are not like, or among the animals; and why we are created in the image of GOD. At our initial conception, we are an unexercised, unlearned, measured probability of desire, called a new eternal soul, gifted out from FATHER'S *creative* soul. We are capable of knowing and being known by GOD, gifted in loving ability and hope, far above all lowly ungodly creatures. Yet most souls fall to be one with the ungodly predatory creatures below. This world identifies mankind as animals, and their familial affinity for animals grows by the moment. But we are meant to grow to be GOD'S beloved eternal Children. FATHER created us in his desire to teach us to be lesser gods, in one mind under and with him. Which means we *can* and *must* learn and treasure our portion of perfection and omniscience with GOD. We are made to learn and know our portions of pure, yet partial perfection and omniscience; which can contain *no* imperfections or false aspects. There is that completeness of all perfection and omniscience held by FATHER. Then from FATHER, down to the least of his loving sons and daughters, we see his created ordered measures of perfect omniscience in those who desire these things; those found and loved by GOD. If all that I choose to know, believe, and live, is only what GOD has taught me, and reject all decent. Though my understanding can never be the completeness of FATHER, my understanding, being a lesser pure part of his, and in all agreement with FATHER, I am then a tiny sphere of omniscient understanding, within GOD'S far greater sphere of omniscient understanding. All imperfections, reside in all beliefs of understanding outside the sphere of FATHER'S understanding of belief. We each can grow in wisdom and be one within his created truth. And there, we see the formula for eternal love, enters the equation between FATHER and his children. From CHRIST down to the least of us, this formula is present when we see eternal love as a perfectly balanced reciprocity. Being that we truly want to give all that we are and have to GOD, and all souls who are GOD'S, without fail. This is that highest eternal love of GOD. It is best said as being one together in desiring and having a shared belief that perpetuates the circulation of love, life, and light. Eternal life is held when we lovingly understand GOD to the point that we greatly fear losing their love and understanding more than anything else. We all greatly fear losing the love and understanding of those near and dear to us, so, we try to not hurt that love, we strive for greater closeness to sustain and nourish that love. Why would our love and understanding for GOD be different, or less. Our love for GOD must be our most passioned, or suffering love. Do not think you can separate love from understanding and logic, we can never truly love what we don't understand, or in any distant way, love that which is not intimate and personal. I trust and know GOD'S every command guides us towards our finding his loving purpose for eternal life. Eternal life is only found through eternal love, and the fullness of worship is where GOD'S love resides. This love causes us to see this world's lies, and any confused covetous desire for the things that fade. We see all of creation rejecting and dying away from GOD'S eternal ways of life. We see hateful liberal zombie insanity growing day by day. In knowing FATHER is a personage with a vast body and soul, who once was all alone, how can we, lacking that experience, comprehend such an existence. FATHER created an external way for him and us to eternally *not* be alone. Yet by free-willed individuality, mankind seems to choose eternal loneliness, and his creation is now filled with evil, pain, hell, death, and uncertainty. I must repeat something existential and foundational for all to understand reality. I can never fully express the significance this has in ordering and defining our perspective of reality itself. We must know our next true stepping stone in GOD'S ordered path, can never be seen by us, before we accept, understand, and stand firmly upon the prior step; so, we must come at this wisdom from many points, but one path. Consider how beyond all doubt, there is this horizon out in front of us where we find the knowledge to truly know and understand GOD. And so, our ability to understand GOD'S created reality also exists, as all reality is created from GOD'S knowledge. So, this knowing that formed all reality, existed prior to its creation. This full knowledge must then hold the perfect answer to our every question about all that we are and sense; every what, where, when, how, and why. These answers may not be the answers most want, or will believe, but they are the perfect true answers. By all true knowledge existing now, it must have existed complete, unchanged, and fixed, before this reality. But as mankind finds something new to them, they sense an excitation of creating knowledge, as it being something that never existed in them before. And if it literally is, a never before pondered false knowledge, it was created just then. But the *true* knowledge of creation, can never came into existence at *our* first ponderings of it. All true knowledge and its perfect understanding are there in fullness right now, never to be added to, or taken from in any way. The sum of knowledge that founds all that is gone, and all that is now, and all that shall ever be, are the same and can never change. But fallen mankind's endless slow small interpretations of incomplete information, can and must change forever. For us, there is only our time of seeking and acquiring our own believed knowledge that remains. Surely this is undeniable; but is this not just one half of reality's big question. By these principled facts, we must see and accept the other half, being that without all conscious perceptions of it, there is no existence of knowledge. Knowledge, means that which is known by someone. Perceived interpretations of information, are *our* definers of reality, so, these form *our* knowledge of reality. But what is this attribute we call our newly generated patterned force of desire. This force, we souls consciously willfully project to create and indwell our perceptions, to bring our awareness of existing in GOD'S reality. Science knows we hear no actual sounds, beyond our interpretated perceptions. Such auditory experiences, rest completely within our soul. Beyond our inherent awareness of sound, there are only moving waves of silent dark energetic frequencies. There are no real physical colors or brightness, no lines or shades that exist outside our soul's interpretive awareness of them. There are only forced, dark, silent, strange, moving, wondrous particles and fields and wavelengths of energy and mass. And all representations of this great external perception of GOD we each behold, exist only within our soul. There is no real physical feeling of anything or anyone, it is all our perceptions of feeling within our soul. Outside our soul, there are only dark unknowing entangled creations coming into the proximity of, and colliding and physically affecting one another. There are no external scents or tastes; all we sense and perceive is within our own experiential awareness. The *physical* universe, including all physical life and the stars are there, but it is all dark, silent, tasteless, odorless, unfeeling, unknowing, and unaware. By this fact, we see most *understanding and beliefs* are created bi-products of false concepts and constructs of individual awareness. An individual's own understanding of reality, need not align with GOD'S omniscient knowledge of the same for that individual to exist in false beliefs through a limited, temporary physical perspective. Every creative soul, by their desires alone, can come to believe an entire life of gathered fictional fantasy. The term "conscious concept" means no more than every created particle/wave-based perception of belief, formed within each soul through desired awareness. All knowing and belief come into existence at our creation and accepted interpretation of our perceptions. So, a great knowledge paradox arises, being, if all *true* knowledge of reality, existed before mankind ever perceived it, and if all knowledge must be conceptual and consciously perceived within at least one soul for it to exist, who first perceived all the required knowledge that aligned this reality. Knowledge must be stored to continue to exist, so, the lawful understanding that initially aligned this created universe, could not be stored within the created physical aspects that did not yet exist. We see there *must* be a vast omniscient creative forceful soul of awareness going before it all, who knowingly perceived and forcefully aligned the quantum and firm condensed physical creations external to our souls; which we sense and perceive in a lesser secondary way. There is no other logical rational option; yet so many choose an illogical irrational option. This quandary of desired created belief, divides the angels and mankind alike into two groups. Please see, that by believing one Supreme Eternal Forceful Conscious Soul exists, every quandary of reason is answered in every way by seeing FATHER'S forceful creations are just created perceptions. In this, all things become clear, as FATHER becomes the first scientific *necessity*. If you deny the truth, all quandaries and enigmas forever remain, to baffle, confuse, and drive your soul insane with endless unanswerable questions. Most are doomed to never know the truth, and by insanity they will never even know that they never wanted to know the truth. Remember, one path answers every question for all eternity, all other paths question every answer for all time. Many hope that in not knowing, there is no accountability; and in innocently not knowing, and never being told, this can be true. But there is full accountability in avoiding, resisting, denying, and *not* wanting to know GOD'S truth. We need to take many steps away from mankind's temporary hopes and belief in a purely physical reality, and towards a consciously and spiritually reasoned and purposed reality. As we each ponder how we experience our growing awareness of reality, we must ask ourselves, is our learning ability and capacity potential finite, or infinite. How can we learn and learn, and then suddenly reach a capacity, and be strangely unable to learn the next simple and relative thing? In knowing there is one Eternal Supreme Conscious Soul who created all reality; and seeing our souls are gifted minute portions of his soul, we must be *consciously* eternal. But we are not gifted with an owned, self-sustained, physical eternity within ourselves, as is GOD. Consider something quite wondrous; imagine the result that must come within your soul at knowing the true foundations of reality, by learning just our adequate portion of GOD'S knowledge and understanding. In understanding *only* things that are eternally true, there must come something very new and different, from what comes by falsely interpreting and believing little temporary pieces of physical data to build a false purely physical model. Such physical beliefs, in adding each new small piece, causes mankind to forever speculate further. But in receiving our pure and adequate portion of GOD'S omniscience, there comes our exacting belief and understanding of all that has occurred to this point. Yet only FATHER'S fullness of omniscience, knows and reveals when and what he brings next. By our adequate measure of Godly understanding, we sense wondrous new spiritual properties; the greatest being the one final truth of why. At adjoining our soul to GOD'S perfect understanding, there comes an inherent disproving of prior false beliefs, that eliminate all speculation about how, why, and when. It is a true wonder to consider the new spiritual elements that arise as we bond with GOD'S omniscience, bringing our true conscious metamorphosis, being key to our inheriting the mind of GOD. Visualize your once worldly awareness as a spherical space, filled to represent your lost understanding with points of dim light and points of pure darkness. These being your imperfect partial rights, complete wrongs, and all desired false beliefs. Picture losing *all* of that, and learning only pure and true things. Building and living in your portion of GOD'S truth is *our* perfecting. Picture your sphere now filled with only the qualities and quantities of the ruling pure truth of GOD'S reality. By this, what we once believed in confusion and darkness of physical desires, turns to GOD'S blinding brightness, receiving his pure truth of all things. At finding GOD'S direction, truth, and omniscience, comes our understanding of GOD'S mind. Isaiah 55:8-9, "*FOR MY THOUGHTS ARE NOT YOUR THOUGHTS, NEITHER ARE YOUR WAYS MY WAYS, SAITH THE* ***LORD****. FOR AS THE HEAVENS ARE HIGHER THAN THE EARTH, SO ARE MY WAYS HIGHER THAN YOUR WAYS, AND MY THOUGHTS THAN YOUR THOUGHTS.*" This world's evil choices, ways, beliefs, and devolving understanding, all arise through its ever-changing core physical desires. And such imperfect worldly desires, are all most souls now hold and suffer under. But in turning fully to the omniscient ways of GOD, these can change and dissolve from our conscious manner of being. If we continue in this Godly process, all speculation, doubt, and confusion disappear. The one true science and religion become pure and set as *one* in a true immutable consonance, being immovable and everlasting. The desire for the confused false answers mankind searches for, no longer exist in us. We hold a firm knowing of what brings eternal life, and all that takes it away. Once we see FATHER'S omniscience and creative force and creation's purpose, we know these *must* forever be the foundation of all shared external reality. As we grow to never doubt or be confused about the conscious *and* physical realities, we never again wonder about the existential truth, and we know every way of life. It is loving peaceful and assuring to see the model of FATHER'S created external physical reality. Every physical question is answered, but the eternal spiritual question as to our soul being there in GOD'S New Heaven, has not yet been answered. I have never found one physical or spiritual problem or paradox in GOD'S model; yet they plague fallen mankind's too slow, too small, standard model. When viewed from GOD'S perspective, both sides of reality align perfectly. And in holding firm to GOD'S omniscient understanding, we find our greatest want is to tear down our temporary self, and forge our soul as a new eternal spiritual quality. In receiving a measure of omniscient understanding, our considerations, desires, and plans, remain ours, yet they become like those of GOD, being of an eternal enduring fashion. We eternally accept and love *only* that which is perfect and will serve GOD'S Family. Consider how all of fallen mankind's intended or unintended divisive thoughts and actions arise through the imperfect desires that permeate their beliefs and understanding. All misunderstandings are born of differing false knowledge, pride, fear, and all individual temporary desires and beliefs. In the opposite fashion of fallen mankind's scattered divisive *individual* ways, GOD'S Children share one omniscient belief, and we have all eternal provision through GOD in one great eternal purpose and desire. FATHER is *in*capable of unintentional or intentional divisive desires arising in his soul, by having no place or foundation of fear, weakness, or evil physical desires, from which all imperfect misunderstanding arise. Under GOD'S conscious primacy of spiritual creation, we see the original truthful reasoned purpose for his creations must always stand and remain, or that same creation cannot. This too, comes by the eternal perspective of understanding and truth in GOD'S highest omniscience. And again, there we find the first to the last reasoned purpose and action of GOD. Those who believe our awareness and life are finite and temporary, hold endless devolving understandings of reality, bringing endless conflict. Consider that in those who do *not* desire GOD'S one eternal defining pure truth, and *den*y their soul is an eternal force; we see their lives ruled by one after another godless dying beliefs, believing in grave error that their literal conscious annihilation comes closer and faster. And as their individual beliefs destroy all spiritual life, they lose all commonality, as all the good shared sustaining things of GOD are lost to them. Are you among the fewer, or the many? Which do you desire? We see the planned ordered creations of physical life were designed and forcefully created by FATHER'S conscious desire. We see his original formula was for forming only good eternal connections to find and sustain eternal free-willed love. And yet, for true love to exist, he had to allow all free-willed choice for hateful separations. For the reality of Heaven to exist, there must be both FATHER'S strongest projections of forceful will, and our aligned projections of all lesser forceful free-will souls, having one belief of eternal purpose together. Did not GOD'S garden exist in perfection, until the conscious alignment between GOD and Adam was broken? From the beginning of all that physically exists, there has been the requirement of dynamic interacting physical layers of the entangled oscillations, rotations, and orbits, that bind GOD'S physical creation. And it being coupled to all created mobile free-willed souls, to allow all desired acts of spiritual and physical choice to arise and exist among the increasing probabilities of physical and spiritual chance occurrence, introduced by Adam's first evil choice. The weaker Eve ate first and tempted Adam to eat. So yes, sin entered Adam's reality through Eve; but it was Adam who held all dominion, so, Adam's sin caused the fall of the physical creation. FATHER'S sustained projected patterns of force, exist over the consistency of all physical aspects; a design which limits the creative actions of rebellious souls. And now each rebel tries to satisfy their every desire for this world. Our choices will always require our physical sensing and our subsequent desires, being our physical attractions and repulsions. Our soul's size and set physical free-willed boundaries allow, yet limit our inner creations and external actions. All physical acts we engage in, are a trying of our faithful responsible desire. If a soul denies FATHER, and their soul's quantum forceful existence, their perspective flips upside-down and inside-out, as they fall away from GOD. This causes the firmer physical side of reality, to appear to be more complex, and the more-vast complex quantum side to appear far more-simple. To materialist reductionist, the physical appears more complex, not seeing a forceful conscious determinism projected from the vast, more complex quantum heavenly side of reality. By their slow small purely physical temporary reductionist perspective, FATHER'S omniscience and the points of his body that form all certainty, are blurred and lost to them within those quantum waves they can't truly define. How can one believe they can build a true model of reality, without first seeing and knowing the very foundation of that reality. They don't know the qualities, or the creative cause for the vast fields they believe are reality's foundation. They refuse to see their understanding and standard model of just a sliver of the physical complexity, that is GOD'S creation, can never be enough to yield true knowledge. They exist in a downward spiral of shrinking expectations, as they fall and diminish, yet they can't see or know this. As materialist find the deeper ethereal layers of reality beyond their touch and knowing, to them these layers appear less complex. Yet the entire understanding of all reality, rests within and upon each pure physical singularity they can never see, set just there, beyond their limited abilities. Mankind sees *their* perceived horizon of reality, keeps moving away from them; yet this does not teach them that the truth, residing at its far edges, can never be within their reach, leaving them forever blind. Out beyond their perspective limits, is a place of never-ending complex binding and dismantling, a line where the foundational aspects of each realm move into the other. Two realms of life, now divided by gifted conscious size, laws, and relative physical perspectives. And by no more than their religious false faith and hope in a false founding physical miracle, Materialist fight all true conscious considerations. They know that allowing any truth into their beliefs, requires admitting GOD is there. So, by their rebellion in pride, envy, and ignorance, while residing in physical limitations, we see a great conscious divide opens. And we see all Materialist desire and choose their ever-failing theories, filled with miracles, unseen variables, paradoxes, perturbations, and zero evidence that shows our conscious awareness is a physical emergent. Is this *not* the strangest weakest godless religion you have ever seen. The religion of Materialism is *literally* the weakest construct of desire and faith that exists in this world, which many hold tight to in a death grip. They cling to it, even as a lot of new evidence conflicts with their desires for it to be so. Their beliefs are part of a humanistic religion called the spirit of antichrist. The spirit of antichrist, includes *all* beliefs and understandings, from the outright denial of CHRIST, who is the first gifted soul and the only begotten SON, all the way down to include each individual belief that alters one word of Scripture. If you do not know the attributes of CHRIST, but say you believe *in* him, you cannot from your own worldly desires, assign any attributes of belief, or ways of life to CHRIST. If you refuse or fail to let CHRIST reveal his own Holy nature and love to you, and lovingly accept them. And instead, take it upon yourself to decide what you desire to believe the loving nature of CHRIST must be, assigning your own desires upon him. This too, is the spirit of antichrist in it being wrong, it is anti, or against who CHRIST truly is. CHRIST desires to be known and loved for who he truly is, being the real true spiritual intimacy of love. CHRIST, nor anyone, wants to be known and loved through another's made-up fantasy of who they might be. Would you want that? I know I can only be *truly* and *eternally* loved by another who knows who I truly am, and not loved for what they imagine me to be. Yield to

HOLYSPIRIT in your spiritual studies to hear and learn of GOD, being the greatest treasure that exists, or yield to the ways of yourself and this world to form your desires and beliefs. These are the only two ways. One can easily fool one's self for their entire life, and many do; but not one soul ever fools GOD for the twinkling of an eye. GOD'S love, patience, and mercy, are far reaching, as long as they are not misused or lied about. We must study and come to know and love GOD for who they truly *are*, and know and trust they loves us back. And at the last, we will find our good eternal home. All who truly love GOD, always emulate CHRIST. The purpose of *our* eternal perspective is to responsibly perpetuate all Godly life in our eternal courses. But the focus of the imperfect temporary fallen slow small perspective, has devolved into prideful envy and self-need; and even further to kill or be killed. All higher point/particle complexities of temporary physical life, are now being over-run by entropy and predation. Entropy exists and increases in a creation, when contrary forces weaken its creative force; even the creative forces that align reality for the purpose of sustaining life. Falling from perfection into entropy, weakens and dissolves all aligned complexity, back down into its original simplest grid state of the spherical singularities that form the vast sphere of the aether. But why does this power of entropy stop there, and have no power over the aether? Because FATHER sets and controls the boundaries for all destruction and dissolution. Let's consider our sensed rich fullness of this current external reality, which is at first a wondrous mystery for each soul to perceive. At first, and on their own, mankind can never understand what their awareness of the physical side of reality is, or where it comes from, or where it can go. But by a pure desire, we can learn how rich our spiritual experiential existence is meant to be, and then understand where we are. Then we can see the two final destinations of where we all can or will consciously go. We all consciously exist, but not in this physical world, or in *any* external physical world. We dually exist, but only within our own consciously created estate. Both our soul's awareness, and all external physical complexity, exist to us, only by the aspects of our created perceptions. The body is just a vehicle filled with sensors, traveling through what FATHER has perceived for us to spiritually exist within; the body is for moving through and sensing the external reality, but we are in no way this body. GOD has all loving reasons to create our perfect pleasing eternal physical environment, but minus aligned spiritual understanding, belief, and agreement, it all becomes temporary. In the Garden, as Adam fell under his individual understanding of belief, the whole of the earth, that had been set under him, fell. Creation did not fall when he ate that fruit, it fell when he made his firm choice to do so. It did not fall at his temptation to do it; it fell when his heart and soul turned from GOD. His eating that fruit, was just the first physical evidence of his fall. The entire earth, there and then, fell from GOD'S perfect purpose; and all entropy began right there with Adam's first imperfection, radiating outward, introducing an imbalance and disorder into FATHER'S sustaining forces. We see an increasingly corrupt and predatory process began, and within just one generation of life, we see the first murder appeared. Mankind now prefers their descending physical understanding and beliefs. They fight for control of this world through all their outward expressions and actions, and have made endless choices, so far beyond what is good. The world is now filled with individual arrogance, as mankind seeks to experience all that is in it, in unfaithful irresponsible insane abandon. They build individual physical kingdoms of their own in being confused towards love, and life's purpose. This creation was not designed to be this way, and it cannot and will not last long. By FATHER, there is coming the last moment of it all, at the harvest of that very last good eternal soul. There is then GOD'S final separation of the lost, who forever fall in their individual created pits of confused hell and death. Now, the more a soul sees the body as who they are, the more they see all bodies as who others are. This is what animals do, having no eternal created purpose with GOD. Since the fall, individual beliefs have pulled lost souls towards their never-ending end. They no longer align with GOD'S, righteous lively understanding, they align only with their self-righteous understanding. This is the formula for a prison of ever-devolving individual belief, being aware of nothing but ever-changing unfulfilled desires. All beliefs in a temporary devolving reality, are born of a very fractured imaginations, who look past every word and line established by GOD. Isaiah 5:13-14, *"THEREFORE MY PEOPLE ARE GONE INTO CAPTIVITY, BECAUSE THEY HAVE NO KNOWLEDGE: AND THEIR HONOURABLE MEN ARE FAMISHED, AND THEIR MULTITUDE DRIED UP WITH THIRST. THEREFORE HELL HATH ENLARGED*

HERSELF, AND OPENED HER MOUTH WITHOUT MEASURE: AND THEIR GLORY, AND THEIR MULTITUDE, AND THEIR POMP, AND HE THAT REJOICETH, SHALL DESCEND INTO IT". These are not physically famished and thirsty, having no bread or water; these thirst and starve in their confused absence of the shared life-giving beliefs, ways, and truth, of GOD. These endless individual beliefs that form the confusion of hell itself, fill those being spiritually dead. If you are eternally lost, your understanding and beliefs will forever change, just as they do now. You will never find an anchoring spiritual rest of knowing the eternal truth, as endless changing lies are what the lost have chosen. The only truth that *does* exist, they have rejected in their desire to be a god over their selves. Most souls take to themselves all latitude, as they want *their* endless divisive individual desires to be accepted among the created eternal ways of GOD. These endless contradictions of increasing disorder, found in all individuals, bring a final fissionable physical runaway event. This universe is beyond reversal, and heading towards its critical mass of utter dissolution. FATHER holds his Holy eternal *experiential* understanding, we do not, and GOD'S reciprocal sacrificial love in *this* world, has become very rare. We must know we are eternal souls, and understand why and how we may or may, or may not, be found among GOD'S family. Consider all you know and love in this world: your parents, spouse, siblings, children, friends, neighbors, and physical realities and experiences. Consider your every tiny effect upon all of them, and their eternal come. If GOD asked you, what would you lovingly give up and always do in every moment of your eternal existence to ensure those you love find eternal life, love, and joy; what would you be willing to truly do. Our loving responsibility is to try with all we are, to not destroy eternal love and life itself, and yet those we love may despise and reject our work. Yet most living by their own individual desires, will say we are judging and rejecting *them*. Most souls believe no one can know GOD *or* their ways. Such a defensive belief and perspective can be impossible to break through. My truest spiritual fear, back in 1974 after I turned back to GOD, was of my misinterpreting GOD'S words, and believing a lie. And I literally prayed over many months, asking HOLYSPIRIT to not ever let me believe a single lie about GOD'S truth. I was so weary of everything and everyone beyond the Holy Bible, I refused to read anything except the Original King James Bible. And since then, I've seen and shared the ongoing pain in GOD'S heart, seeing most souls drowning in their rebellion to their full truth. Now if you have, or do surrender all; most others will see your Godly belief as being of and originating within yourself. But never give up hope, some may sense a bit of truth to grow within them. Until you see only two spiritual families, being eternally above your fallen physical temporary family, your familial worldly love will have a great pull on you. Most spend their lives *physically* defining reality, trying to deny and forget they have any eternal spiritual responsibility for each other. GOD has promised, that if we faithfully dedicate ourself to join their common loving effort, they will tell us every perfect word of wisdom that we will let ourself hear. GOD helps in every Holy Righteous step that we can and will see, and towards every perfection we can desire. Our hearing and living this way, is the purest love we can ever declare towards GOD, and all souls. Will we crucify our selfish desires and pleasures in the rich firm life, which is but a blink of your eye in eternity. Can you devote your Spiritual love to live your unfolding moment, emulating CHRIST'S pure example of life, before all who can love, as our KING did without fail. Or will you seek and enjoy the imperfect temporary pleasures of this temporary reality, until they soon suddenly end. I ask you, to what true extent does anything temporary, literally exist. The longer something exists, the more it is real and true. So, we see a great difference between a temporary provision and the eternal. In our eternal life, is a temporary physical experience any more real, or worth any more, than a simple dream of doing the same. One soul may be able to physically and temporarily do something others cannot. But others can imagine and dream of doing that same temporary physical thing. All experiential reality, exist in our soul's created perceptions. So, after a temporary, actual or fantasized experience has passed, do we keep anything more than the thought from each. Is it of some grand eternal importance to say we once physically experienced this, or that. In all reality, what eternal physical aspect, or truth, do we take away from physical experiences that we can have, hold, and cherish *forever*. Love resides on the spiritual side alone, and eternal love resides in a pure spiritual way of life. Our desired understanding, beliefs, and ways of life, are the only formed elements of our soul we take with us. Do you desire and need spiritual life, or do you desire and need those physical experiences with others. Physical desire, is all there is in the life of the animals below us; they cannot spiritually aspire. We must sense the complete difference between the joys of spiritual, and physical experiences; between a Godly and animal experience. Study all animals to see if they know of GOD; they do not. Godly souls see the elements of eternal love in sacrifice, hope, devotion, responsibility, faithfulness, and the fear of losing what we love. This highest love, when realized and protected by each receiver, brings eternal life. All souls know how to physically give here and there in their moments; but most cannot equate this to any eternal spiritual sharing or giving of life's force itself. GOD'S commanded ways of living, are for holding us all together as one life with him eternally. What type of strong bond do you imagine is required to hold us together for eternity. Most believe that in life, we can be generous and then selfish, and then generous again, and so on. And in this they find a satisfying blended physical and spiritual life. How can any temporary physical way of life appear as a higher form of love, to the lost. Most know deep down in their soul, there is no true eternal joy in fulfilling any physical desire or need. Yet most don't live or venture deep down, they live mostly at their shallow desired surface of their soul where providing eternal spiritual love, life, and safety to others means nothing. A weak believer in GOD, exercises many powerful physical aspects in their way of life; and so religiously, they are still part of the physical nature of the fallen flesh of mankind, this is an instinct most refuse to overcome. And there are also those who do know deep down that the way to over-come all evil, is to lovingly emulate CHRIST, yet never go down that path. There are so many worldly desires that keep souls in bondage and a great distance from GOD. CHRIST came to give us forgiveness and a reason to overcome, escape, and strive be set free of all sin. A loving FATHER knows no greater joy than giving all they have and can to their child; yet most earthly parents don't see how most of what they give is not for their child's eternal good; but for exercising and enhancing their child's physical temporary desires. In most cases, all their giving of abundant physical provision, is spiritually blinding and destructive to a soul's Godly understanding. But as long as a child just *says* they believe in a god, a weak believing parent is satisfied, as this is the weak standard most apply to their self. We must see that if our practice of GOD'S ways, does not bring us joy in this world, that practice would not bring us joy in Heaven, where GOD'S eternal ways must be our souls great desire. Our desire for GOD'S ways must cover our surface, it must fill our core, and everything in between. Most souls on their day of judgment, will wail at not being brought into Heaven. These cannot see that the true peace of Heaven is a *way* of eternal shared love, they do not have, or even desire, and never will. They cannot see that the eternal ways of Heaven to them, would be another hell they must try to change to suite them physically, as they are so carnally minded (physically minded). Right now, we are building our chosen *spiritual* foundation and structure of Heaven or Hell deep within us. The Kingdom of Heaven, born of the mind of GOD, is revealed and resides deep within, and covers the surface of his Children's souls. But the physical things and ways, cause individuals to fill themselves with hell, as they are set adrift and alone forever. Individual confusion fills most souls with untenable desires and cynicism, as their hell stays and grows within them wherever they may be. Again, if GOD revealed every perfect wisdom to you, for you to live a life aligned with eternal joy and peace, is this something you truly desire? And why? The many, who never have this free-will desire, see that is an impossible demand of being forced into a non-physical, spiritual way of life that does not allow their core physical desires. Luke 13:23-24, "*THEN SAID ONE UNTO HIM, LORD, ARE THERE FEW THAT BE SAVED? AND HE SAID UNTO THEM, STRIVE TO ENTER IN AT THE STRAIT GATE: FOR MANY, I SAY UNTO YOU, WILL SEEK TO ENTER IN, AND SHALL NOT BE ABLE*." Why would many strive (try very hard) to enter Heaven's straight and narrow gate, and not be able to do so? These may have a strong belief God is there, and hold a certain measure of desire for GOD; but their desired ways of life, do not fit through GOD'S Gate. Nor can they understand how anyone else could desire to live in a spiritual way above a physical way. We who lack the physical desires for this fallen world, and have an eternal desire and perspective for reality, see the physical desires of the lost, as an aberration, having no eternal value. Now if *this* core understanding is, or becomes yours, you will only long and search for the eternal things of GOD. At our true turning and repenting, we *strive* to separate ourselves from all ruling physical ways. Heaven is not about forcing ourself to follow rules we don't agree with, for eternity, it's about freely becoming one with GOD in belief. It is the joy of finding that in all the eternal ways of love, we are like GOD. Heaven is for souls that love and emulate who GOD is; and if any do not know and love these things above all that is in the world, Heaven is not within them, and can never be Heaven to them. Yes, in many, there is a lesser imperfect measure of temporary Godly love. Which again, raises the Judas love equation. It is hard for us to imagine that our love for GOD could be insufficient and fade, making it a temporary love. But we are sure Lucifer, Judas, and others, never considered they could lose their love for GOD. By GOD gifting, creating, and allowing all free-will choice and true physical temptation, we see destruction of spiritual love, spreads to all weak souls. All evil things must be allowed during our time of choosing, but at the end of grace, all evil allowance must and will be stopped forever. The greatest requirement for any and all love to exist, is in our gifted allowance of free-will desire and expression; and the second is a created true free-will choice. If free-will is gifted and required to allow and raise the rewards of love, then all hate, denial, abuse, and pain of loss, are the cost of that love. All rebellion must be allowed, but rejected, for sacrificial love for GOD to be real. As FATHER gifted the first 13 souls, rebellion did not appear until Lucifer the 14th.

But with each new soul, uncertainty was added to reality, as evil became a certainty. Now the embodied Antichrist will be a fallen angel appearing as a human's body, who was the 666th soul FATHER gifted, and the second to sin, out of all the numbered named souls in Heaven, Rev. 13:18. Scripture does not explicitly say this, but we see rebellion never arose in CHRIST, HOLYSPIRIT, the 7 Arch Angels, or the 4 Seraphim. We also know Lucifer, the 14th gifted soul did not quickly rebel, we don't know how many years of lively heavenly experience Lucifer had before sinning. We do see sin first appeared a certain distance from GOD in the order of gifted souls. We know there were *no* pre-set influences placed in the lesser and lesser order of souls gifted from FATHER; but their bodies were nuanced and different to certain extents. Scripture says Lucifer was broadly loved and beautiful, and came to see himself above those in the Holy of Holies, for his physical beauty. Each new soul is set at the center of true choice. And free-will means GOD created, but forces no choices. And now mankind, are born into a fallen physical world of sin. And from Adam and Eve on, we all inherited increasing bodily weakness and tendencies towards this physical world. We are born into a world of physical sin, surrounded and tempted by it all. So, FATHER offered many counterbalances and remedies to correct and restore us. The proof that free-will choice exists, is seen in all probabilities for souls to either want or reject GOD'S ways, are now evident. And the *cost* of each newly found, rarer and rarer, free-will love among each new rebellious generation of souls, is greater. This Law and Principle of an unfolding *greater* cost for *fewer* souls, by diminishing returns of free-will love, is now part of this shared free-will reality. GOD brought all external realities, which set and allow all choice, chance, eternal love, and every antithesis, starting at that first great moving forceful love between FATHER and JESUS. And as we see the desires all souls hold, and the choices all souls make, we see every form of confusion and opposition that can be imagined, has been desired, believed, and lived by some. Every weak incomplete false conscious construct of belief, can be accepted by someone. Now in that love and hate are antithetical forces; what might be the force antithesis of fear? It is not joy, wonder, or elation, as some may think, it is hope. Fear and hope, are the antithetical emotions we sense when considering, not any former or current moment, but the as yet probabilities, we anticipate in our unfolding life. The emotions of fear and hope apply to what may yet occur in our expanding moment; just like regret and shame are associated only with former loss or failure. And we see love, sadness, and anger can be associated with the former and present moment; but again, hope and fear are associated only with all our as yet coming moments. Is it a strange thing that one person can fear a certain thing happening, and another wants it. One person can have hope in a certain thing happening, and another have no such hope. One can fear the thing hoped for by another. It is our core beliefs that give rise to our fearful and hopeful emotions. Some souls have fears or hopes that others see as irrational, being beyond the ability of them to hold. Some strong fears are called phobias, but this does not make the fear unreal, or irrational. Some people have many more strange things happen in their lives than an average person. Is this fact itself irrational? If strange occurrences are more common for some, why should their fear or hope towards strange occurrences be irrational? I have had *many* close calls with physical death, and many other inexplicable things happen in my life, far more than what I hear is the common experience. By experiencing many inexplicable things, I can't see tempting fate by doing certain things, like getting on an airplane unless GOD moves me to. If a consensus says this is irrational, it is irrational to them, according to their experience; yet to me, it is the sane thing to do. If something typically happens to one in a million souls; but I sense a far higher probability for myself, I will hold a greater fear or hope towards the rare occurrence. Let's examine many souls differing levels of fear or hope towards GOD. Again, by my high number of super-natural non-physical and physical experiences, I know many of my experiences came by seeking GOD for most of my life. Over 100 inexplicable other-worldly anomalies and exceptions to reality, cannot occur in 60 years, to one random soul having no neurological imbalances or deficiencies. So, *I* know many experiences had a message and meaning, and came for a reason. From conception in the womb, GOD shares spiritual gifts to open our minds to his existence; but rebellion starts early in most souls, preferring warm milk, blankets, gentle caressing, and a soothing voice. Rebellion, being denial, can start within the womb, or at any time, and grow or wane. And it can be overcome completely by those having the desire and will. GOD speaks louder to, and teaches those who cry for him to speak louder and teach them. GOD is the greatest treasure to be found; so, what has been your determined effort to find the fullness of such a treasure, do your desires and efforts still grow towards GOD. Growth is seen between our past and current love for GOD, and as we learn, grow, and fill our hearts with them, they turn and look upon us in hope and wonder. When we come to sense that GOD truly sees us, it is wondrous, yet fearful. GOD, in seeing any hope of what they search for in us, they begin to nurture and come closer to us as we grow in our eternal loving responsible faithfulness. Once after a certain lesser sin, I was repenting until I felt sure, I had regained my strength against that desire before formally asking for forgiveness. But after a week, I felt HOLYSPIRIT move within. Had my forgiveness come without a formal ask? Does repentance require we formally ask for forgiveness? We know confession, being admission, is a true part of our shame; but what constitutes admission in our heart and soul? We know repentance is a growth process, just as with all ways of love and lively belief; it is not the flipping of a switch. So, if we fall behind to pursue a worldly desire, our return to GOD *can* be a just as long, or longer corrective path. Scripture says the body we are adjoined to, is to be kept as a Temple of Righteousness and Holiness and one with GOD. Repent and pray for HOLYSPIRIT to come and dwell with you. Such indwelling will occur more and more, as we more and more, keep the body a Temple of GOD. If HOLYSPIRIT dwells within us, he is *master* of the house, and *we* are a guest. If we defile the temple, we offend the ways and mind of GOD. If we walk and live only in the presence of GOD above all things, we hold our greatest love and fear of GOD. But if we walk and live out of the presence of GOD, this love and fear have faded. To those who only sense and experience the physical side of reality, we appear foolish, superstitious, ignorant, and maybe insane. But how are we ignorant or insane in knowing the ways of GOD, that others simply refuse to know? Who are the truly insane, us or them? I raise the issue of the fear of GOD, because there are two great emotional mechanisms that lead us to hold onto GOD. There is first our love for GOD and the ways of eternal life; and there is our fear of losing GOD. Do not let a worldly Preacher tell you, you can never lose GOD. You can lose GOD just as fast as you found them. The closer and more real GOD becomes to us, the more we feel our love and fear towards them. If we do not know and feel both, we can't truly know or feel either. If you think you love anyone, but don't fear losing them, that is not love. Most souls can't understand why eternal love is defined by GOD alone. One can form a belief in *a* false god who they believe will satisfy their every physical desire in a grand eternal life after death. One can form an unaccountable belief in a false god, that lacks fear of losing that false god. But intimately knowing and loving the one true GOD, is found in our loving fear, to emulate and please CHRIST. I have grown to intimately know, understand, and eternally love GOD, over decades of true desire. I have awakened to pray, ask, sit, and write about these things *every* single morning, for decades now. I never tire, and my desire never fades. I sense a greater measure and strength in these wonders, year by year. I still spiritually grow without restraint or limit, as GOD has called us to do. To love GOD, we *must* love their pure righteous holy ways, desperately wanting all things of GOD in who we are; and fearing our loss of these same wonders. A true and righteous love for GOD causes us to see that this world and all that is in it, has turned in opposition to GOD. Individual beliefs in *a* god are strange; all such beliefs are based upon individual desires for their god to be seen in the mirror. This world is filled with endless waring false beliefs to choose from. Just want it, and believe it is ok to want it, and you have formed your own personal religion. This world's fallen godless ways of life, are nothing like FATHER once made life to be. A false religious perspective, believes GOD made all physical things to sustain and make us all happy in our individual physical desirous lives. But our love *for* GOD cannot be real, outside our desire for reality to be and remain as FATHER made it to be, to please *him*. In all these things, FATHER is good, righteous, holy, and merciful. The good words of GOD will forever secure our enduring reality in every eternal way. This we call a top-down conscious formula, or a formula that sustains all that is good, coming from GOD above. There are now endless false bottom-up formulas, known as individual interpretations of knowledge, false politics, false science, and false religion. Such souls resist receiving and living the top-down created truth of GOD. Why do so many believers think they can reject the omniscient created ways of GOD? And not see it as then rejecting GOD, and raising a paradox. Why do many souls err greatly about our awareness, and our created perceptions? With no true reason or purpose, many accept our perceptions are formed by brain meat, being the end-all essence of their conscious awareness. I can't see how to begin to accept such a slow small illogical inexplicable belief. I know when I visualize, the image I forcefully create within my soul, is a required physical corollary aspect; but my awareness of it, comes by an other-worldly, unique, and distinct quality. FATHER divided yours and my portion out of his own inherent intrinsic phenomenon of creative self-awareness, in his hope we would desire and come to be his loving Children. All loving souls desire spiritual children, we hold this innate inherent desire, because we are created in the image of GOD; but a desire for children, dies in souls having no eternal purpose. It is a strange-phenomena, that as true understanding of reality is rejected, souls lost in lies, think their ability to know reality has grown. A mind that loses the eternal defining truth of GOD, at first, develops an unleashed feeling in a vast search for self-truth within their desires. GOD'S truth of reality is surely defined, having its spherical boundary, sum, size, and fullness within FATHER. Take FATHER'S defining truth away, and the search for endless alternate truths implodes inward and downward, as hell's confusion grows in souls without measure. Mankind debates fantasies outside FATHER'S created boundary. Is their debate within their eternal soul, or a dying physical neural system? Pure Soul Dualism requires both together, as both do exist. Many accept a physical system of awareness, with no empirical evidence to support that belief. Has GOD hidden their truth, to leave us blind and stumbling about; or have many souls by desired choice, blinded themselves to GOD'S truth. Choosing to deny the existence and rule of GOD, makes souls individuals, and allows all temporary moral latitude and physical freedom in this world. Most souls love the physical ways and things mankind has made life about, with little to no time or desire to consider the ways and things GOD has made life about; and even less time to emulate them. From a temporary perspective, life is about the physical experiential moment. But from our eternal perspective, life here is surrounded by evil, and about being led home to GOD. In their degrees, most souls are lost in desires to chase, sense, and experience, their created worldly perceptions and satisfy their chosen way of life. Again, if this body senses the physical information of a pencil, my created perception *of* that pencil, was not one occurrence; and my following awareness of it, is a third occurrence. I have heard materialist say they have no desire to disprove GOD, but only seek to prove the physical; such souls live in semantics.

They look for our awareness where it is not, and define it as something it is not. The lost will say, deny, do, and believe, anything they see as remotely plausible, to avoid the truth that haunts them as they hide in their own dark shadows from that brightest light that surrounds us all. Most religions in the world, including modern worldly Christianity, *accept* many errors put out by material science, which turn their soul away from knowing and understanding GOD or his creation. Most can never find GOD'S truth, it being just there beyond their own desired prison of false belief. The religious materialists are lost in a belief that GOD does not need us to know them or their truth in any first-hand purity. Such belief reveals their desire to *not* know GOD in any true way. The lost inherently know deep down, that once they *know* the purity of GOD'S loving standards; it requires their loving eternal adherence. They see GOD'S ways, as superstitious religious extremism. It's unavoidable that in GOD offering love and choice to our lesser souls, that by rebellion you get ever-increasing religions, denominations, and sects. We see how all imperfect thoughts and words pull most souls from the light of GOD, down into individual darkness. Prov. 4:24, *"PUT AWAY FROM YOU A DECEITFUL MOUTH, AND*

PUT PERVERSE LIPS FAR FROM YOU." Separate yourself from speaking evil, and from *hearing* all who do. CHRIST came to *correct* the errors of the Church as an example to emulate; and to pay the pain and passion for our past, *truly* repented sins. But the Scribes and Pharisees could not accept that they too, had been lured from GOD'S true ways. So, *in* their denial of religious-preservation, they declared CHRIST was a blasphemous extremist. These leaders had blinded and deafened themselves to the point they could not be corrected, even by GOD standing before them, and this is the condition of most of today's religious souls. Are you of the true few, I pray you are. No matter how much GOD wants to give to each and every soul, all that is defined as good, they can only give, if we of our own desires, will receive it all as it is, in its true ways. Most all mankind has ever done or said since the fall, was done and said in a lack of understanding reality, not seeing that awareness once filled FATHER alone. How can religious souls believe FATHER did not divide and share his eternal self-awareness that already exists, and believe he created an impossible different form and quality of creative awareness that inexplicably pops out of the wet fatty meat of a brain. Most think we first see, and then we believe. No, our desired beliefs are there first, and then we see and exist in our desired interpretation of reality. They can never prove me wrong with endless suppositions, biased interpretations, and forever changing the meaning of all words. They don't see that when they change the meanings of words, the correct interpretation of history is immediately lost to them. They always fail, because they can't ever prove their own false beliefs. The fallen can only fail more and more, descending into endless insanity, filled with unreachable goals. Only GOD'S omniscient belief can be our eternal anchor, so, adopt it and live. Existence is real, and GOD is the reality that designed and perceived it all. Materialist say believing GOD created all things from nothing is absurd; and believing that would be absurd. But Scripture never says GOD creates anything out of nothing. Scripture says all things external to GOD'S Person, are made real by the forced patterned point-based wave properties that form the substantive breath of GOD'S spoken words. And Materialist are the ones who say that everything can be created from nothing. Can materialists ever see past the physical? Can they see all objective *and* subjective realities? Are they the omniscient authority on all things existential? They claim a superior understanding, so, even FATHER must be envious. Scripture says these dark souls are void of all truth. FATHER perceived precept upon precept, he designed and aligned line upon line, and created all layer upon layered three-dimensional physical certainty from the pure physical singularities that are his body, breath, and words. In their sensing only the largest of all particles, materialist squirm over the evident consciously deterministic process mankind calls the quantum mechanical observers affect. But in all certainty and completeness, FATHER has shared his truth of reality with us his Children; he did not create a new form of conscious awareness, as there is no such thing. There is only one form of creative forceful conscious awareness that exists. And in FATHER'S vast unimaginable measure, he is self-divisible, and can share this wonder to certain extents. And each new gifted free-will soul, has no certainty of outcome, only a hopeful probability of *desiring* that one true omniscient purpose and truth of reality shared from above. We all start out quite blank and unlearned. Adam and Eve's souls, were gifted and adjoined to adult bodies in the daily presence and teaching of JESUS. But we started as souls adjoined to single fertilized cells, surrounded by a fallen creation of rebellious choices for disorder and insanity. Can GOD *ever,* even in Heaven, give us the eternal ownership of a body, making us far lesser self-sustained Gods below him. The first hurdle for that is, can we *ever* learn the fullness of the truth of omniscience that would ardently set and frame us in the *eternal* loving faithful responsibility of oneness towards GOD'S plan. And though GOD may desire this, it seems it might add uncertainty back into his eternal plan. These bodies we now indwell are insignificant, compared to the size of the universe, but they are still quite immense. Consider being six foot tall and composed of around 9 billion, billion, billion, atoms, each composed of many particles, which are each composed of quarks, which are each composed of many truly physical spherical singularities that form the aether, or the breath of GOD. Materialist do not believe in purely material aspects, yet all pure Dualist do. Materialist do not believe in any true consciousness, yet pure Dualist do. Do materialists believe in anything real? No. They cannot see anything eternal or real. I find so many problems with the growing list of what materialists refuse to believe. The sum of their beliefs, reside in unanswerable questions, physical mysticisms, paradoxes, impossible theories, and hell's insanity. Consider why GOD chose the sphere for every created aspect, from our souls to his aether, to particles, to worlds and stars? Because the sphere is of GOD; and the sphere is the only eternal model, and the only reality where its entirety is also its foundation. If your model builds upon a unique defined foundation, we see two aspects arise: the supportive, and the dependent; like a building upon its foundation, or conscious awareness upon the physical. But to try to define a unique foundation of a sphere, you must say it is its very center. But that is a strange fact, as one tiniest common single spherical point, must be the foundation for every size sphere, no matter how massive. So, a sphere has no unique foundation, it is all foundation. But a square, or triangle-based reality, or any other shape, must fail to incorporate its entirety as one. A sphere is the only reality, that can have the eternal conscious and physical prime formation of an eternal singularity. FATHER'S soul is a sphere, and his physical quality and essence that form his body, are perfect spherical physical singularities. A sphere of a single physical quality, is indivisible. Not just conceptually or mathematically, but literally and physically. A singularity has no lesser parts, it is one quality in all ways, containing no secondary or lesser aspects; and it defies any definition of being created by any lesser aspects. In a true singularity, we see the reality of *both,* non-complexity, and a singular pure eternal complexity. Can simplicity be viewed as complex? Can complexity be viewed as simple? Yes, but only in seeing and knowing it is then eternal. All physical complexity and energy now lose their definitions to entropy; so, they are no longer pure or eternal. How can materialist believe that as reality expands and grows cold, that reality must eventually turn into pure space? What will remain? Will space always be something? Can nothing exist in between something? Can there exist such a thing as an absolute zero, energy free, field? Only if you accept pure physical singularities. If you can believe nothing can become everything, you must also believe everything can become nothing. But what and where is its cause or catalyst for non-existence? And would such a cause or catalyst be something? Now FATHER'S and our spherical souls are not bound by any physical laws, being spherical forceful desires. So, we see FATHERS spherical soul is divisible. This vast universe was once a perfect dynamic spherical construct of order, built of layers upon layers, out of pure physical singularities. And now human bodies using every means possible, sense at most, 6% of this creation's complexity. So, by their slower smaller perspectives being set at the conscious divide of reality, we know they can never sense the physical reality's far lesser and far greater ways, qualities, and quantities. Yet their slow small understanding seems to be in awe of how small or large and slow or fast the parts of reality are, that they do sense. Human bodies are quite immense creations within the universe, yet the angel's ethereal bodies are vastly more immense than ours. And they too, are consciously bound and founded out from the conscious divide of all that can be sensed and created, or perceived, as is GOD also. By this order, FATHER, then his SON, then HOLYSPIRIT, then all the Angels in their created orders, sense and know more and more of what has been created in both the smaller and larger directions, than mankind. Why do the godless reject the fact, that beyond all chance occurrence, the human body is designed solely to be moved within, and sense the external creations of GOD, to enable our soul's awareness? Why is this? How could a purely physical thing, no matter its complexity, sense and know another purely physical thing? The physical sciences can never describe or define what it means for a physical thing to know other physical things in a true awareness. How could anything being purely physical, hold or have any self-aware, self-animated desire, towards an external reality? How can any purely physical thing, have self-aware desires, questions, understandings, or beliefs? All we hear from materialists are very strange, very convenient anthropic non-answers of it just is, because it does exist. The conscious awareness side of reality, is as far from and different from, and as unrelated to the un-knowing unaware physical side of reality, as anything can possibly be. This most basic fact alone, affirms Soul Dualism; and just as strongly denies Materialism. Our believed understanding of GOD, is also as non-physical and other-worldly as anything can be, residing and occurring outside of and beyond all possible lawful physical cause, effect, or correlation. Now false beliefs in GOD can arise from simply questioning the endless problems and doubts that arise in materialism. The eternal soul is a stumbling stone for all who desire all things to be physical in an effort to deny their accountability to our Creator. They can't see, that physical interactions alone, can never cause one tiny aspect of awareness. Reality has always unfolded by forceful conscious desire, projected upon the physical. A purely physical aspect has no generating means, reason, purpose, or desire in its self, to force any aligning interactions. Force has always been a consciously generated patterned entangled aspect. Yet Materialist long to find just one closed physical reality. We must see three conscious, and three physical dimensions of reality knit together, being realities six dimensions. Materialist relentlessly refuse to see a recurring constant in their science, being they can never prove an empirical falsehood, yet they irrationally forever try. They believe what they now physically sense, holds their answers, but it only holds more of their endless questions. I see why their every attempt to reach a sound mathematical beginning fails. They think they are always close, but this is their delusional illusion. They can't see reality's physical boundaries, and the conceptual boundaries of mathematics, are very different, which blurs *their* line between what is true, and what is possible. The scale of conscious force over physical reality, which they do not see, reaches so far beyond the bounds of their comprehension. They see their physical horizon; yet reality's horizons are different, and far beyond. They are like naïve children, hoping their dreams come true. They don't see that their constant failures are a result of their desired belief in a model that is so insufficient at explaining reality. There are no beginnings or ends where they expect to find them. I ask you to observe two souls: one who knows he must in love, emulate GOD in pure truth, or be eternally lost; the other who sees zero accountability beyond this world. Which of these two, can deceive others without fear of eternal accountability? Which has claimed consciousness is physical, without any proof? I sense knives coming out at that statement, for I have said I can prove the existence of GOD. But I actually said, I can prove GOD exists, up to and past the same point of anyone proving any godless theory. I have constantly declared *all* and *every* objective knowledge and its every subjective interpretation, are matters of faith, hope, belief, and perspective. I am honest in providing both sides of all the desired aspects of creation, with all spiritual motives and meanings. What I have said, and what they say, do not equate, qualify, or quantify, in any *eternal* way. Yet they fight to deny our free-will right to teach and disagree with them in the classroom. I *know* they are wrong, and they *hope* I am wrong. I greatly expand the argument to allow *all* observed realities, while they minimize it. They deny our forceful soul, and sense just six percent of the created half of reality through their slow small insufficient 4 dimensional belief, to form their theories. Their limited inadequate biased understanding, could never even begin to define or know the fullness of reality. Their physical beliefs arise from crippling inadequacies, as they refuse to see their constant failures reveal they are on a wrong path. They throw pure illusion after illusion against a wall to see if any might stick by any weak consensus. They have no more than a vain religious faith, hope, and worship, for their own un-provable godless god. They're drawn by, and driven deep into paradoxical unprovable notions like multi-verse, and quantum gravity, showing each theory is as hollow as the vague fantasies held by every other desperate soul. These are conspiracy theories, born of allowing only the narrow limited physical data they sense. The physical sciences find great obstacles and caverns in their understanding, so, they suppose and propose what *might* fill the gaps, or what *might* be on the other side, to allow them to forever search for their abstract illusions. Materialist apply abstract patches, braces, and bridging, made of false hopes and equations that arise within their small slow 4D temporary perspective. And if others see some things the way they see them, by reason of a consensus false faith, they feel affirmed and confirmed. These practices raise hopes, and can *appear* to work up to a certain point. But since 1920 quantum mechanics has turned materialism on its head. They cannot get the physical reality to align with the ethereal quantum conscious reality, so their theories are fatally wounded, stalled, and in decline. They sense this, but cannot admit to the world that they now meander about in stages of denial. And by still denying the conscious influence they see everywhere in everything; it signals the end of this creations purpose is near. GOD will soon find this creation bears no more possible sanity or true love; which is the only reason to allow its evil decline to continue. Fallen mankind is blind, seeing so little in a linear 4D way, instead of a spherical 6D fashion. The true foundation and boundaries of reality are beyond the reach of any allowable technology. Again, these physical senses are not meant to be our focus in searching for what we are. Now if you are driving a car, in a limited way and limited moment, you become one in that where you *will* the car to go, you go. What if you never came out of your car for your whole life? It could become a greater and greater physical part of who you see yourselves to be, if you allowed that to happen. See the body you indwell as a temporary vehicle, and never allow it to become central to who you are. The quantum realm soul must always be known as the aware supreme aspect of life. We must know our soul is that which forcefully orchestrates our created existence. We all perceive reality in outward ways, from one conscious center plane or point of reality. GOD says to observe these two states, and understand his work. In this world, mankind has trust and desire, but in the hereafter, their reality will be utter loneliness as inter-soul connections of communicable belief and understanding are lost. Even now, all love, hate, life, and death, are being seen from increasingly different perspectives. There are only GOD'S true answers to all his true questions, or fallen mankind's ever-changing beliefs and loves that equal ever-failing beliefs and loves. Once we see *how* and *why* the forceful created omniscience of GOD leads to an immutable impenetrable eternal love, we *know* it has no inherent or innate weakness, doubt, or lack. We see our eternal need for the Spiritual Constants and Principles of GOD'S ways and belief. There is nothing near or above FATHER'S forceful omniscience that can come against it. In this, we see it is *impossible* for GOD to make an imperfect choice, or hold any temporary perspective or purpose. All imperfect choices and temporary perspectives are formed by a confused emotional imperfect temporary understanding. FATHER'S eternal reason and purpose for his every choice will be known by his children; and inside GOD'S mercy, there is no hesitation in judgment. FATHER'S omniscient love is a guiding force in all judgment, being an experiential-quality, we cannot hold. Let's consider the conscious-physical concept of freedom. Freedom in essence, is in allowing our ability to live and act out our life according to *our* desires; yet desires come by free-will, not by physical freedom. We note that no one has *ever* held any semblance of complete, unaccountable freedom; as universal freedom equals pure chaos. True peace among many souls, requires love, trust, structure, order, reliance, restraint, and shared beliefs. And eternal peace requires we all see, know, and understand GOD as our true right and required provision. Perpetuation of life, love, peace, and even reality, can never be found in a liberal or total freedom. I can easily show, using Scripture and science, that absolute freedom will never exist for us, or even GOD. FATHER has firm binding ruling beliefs, and his love and joyous personage are truly and eternally subject to those beliefs. FATHER has said he does not and cannot change. But by many *taking* temporary freedom to live contrary to his ways, we see constant change in them, and a desired demand for *their* individual freedom of love and belief to be accepted by all others as being good; meaning they hold themselves above GOD. We see most who say they are believers, reside in a religious state of belief + doubt. Their doubt being equal to their weakness and disobedience. What I mean is that most professed believers, think belief means to hold to a firm possibility that GOD exists, yet think it's okay to have personal reservations and questions about exactly who and what GOD might be. FATHER'S plan is to find *his* strong and true love and belief in others. So, it comes down to, is our freedom completely defined by GOD'S commands, or completely or partially defined by ourselves? Or both? Rebel souls want to be at least partially free from GOD'S rule, yet restrict others from opposing or disagreeing with *their* chosen freedom. No one can eternally love any other soul who forever violates the Godly ways placed upon us all to enable eternal love and peace. In this fallen world, we see souls acquire temporary power, privilege, possessions, and their individual desires and beliefs. And these drive every godless individual, to try to share, spread, and to differing extents, force their forms of love and belief upon others. This is why religious liberty requires a limited, small Federal Government, and a less-limited state, and stronger yet, local governments to protect and preserve communal moral understanding and purpose. The purpose of Federal Government is to militarily and economically preserve a country's sovereign borders, heritage, and its historical founding Constitutional principles and interest. The purpose of local governance over free people, is to protect and preserve the uniting spiritual desires of the local majority first and foremost. State Government is in the middle, for a moderated spiritual and economic peace and protection. And a citizen's self-reliant, self-governance of responsibility and Godly faithful forbearance, are our required standard. Now our right to bear arms is essential to preserve our rights against an oppressive government. Communist, Socialist, Fascist, and Dictators, all hate GOD, self-reliance, self-governance, and any freedom of religion, speech, or a right to bear arms. Evil souls seek to restrict and destroy religious liberty. Communism and all her sisters, are matters of centralized power, filled with elite bureaucrats controlling everything by cold godless statistics. The deep state, bit by bit, steals all free ability to protect and provide for one's self. They want us all to be children of the state; they think the role of government is to be everyone's parent and guardian. We must teach, encourage, enable, and allow, all souls to be responsible and accountable to GOD to the best of their ability. And allow citizens throughout all our nation, states, and localities to govern themselves under their ancient Judeo-Christian beliefs. And requiring all to abide by the U.S. Constitution; it being a Christian document, under which, not all religions can peacefully exist. The free-exercise of Judeo-Christianity, is our Constitution's main purpose, being the truest religious liberty that exists on earth. If anyone in any way, tries to restrict a soul's non-violent Christian's way of life, they are in fact forcing their own ungodly belief upon that soul who has become their victim. As a people, we must incarcerate or banish those who violate laws worthy of such punishments. And being under Godly order and control, we are to welcome all we can safely allow, those being responsible accountable minded souls into our country and life. And we must reject all souls who are irresponsible and forcefully contrary. Without Godly standards being applied, the good and responsible must bear all physical and spiritual hurt and costs brought into their midst by the evil and the irresponsible. We must never forget this world is *filled* to its brim with the enemies of Godly religious liberty who unite contrary peoples together in this purpose. In the U.S., only the true Christian Conservative faction of the Republican party, fights to protect the souls responsible and accountable to GOD and each other. But we lose again and again to the souls who allow, accept, and encourage, all who are irresponsible and unaccountable. Liberal zombie politicians, preachers, and teachers, are irresponsible, and see themselves unaccountable. Now the indoctrination of Godless impossible evolution in schools, has been set as the national standard since 1968. And by the liberal control of the media as a propaganda tool, the U.S. is set on the brink of spiritual destruction. We must see the true picture of reality, being that we were brought forth from out of FATHER, under his eternal intension for us to love learn and live in all his eternal ways. So many struggle with this created truth, wanting both GOD, and their imperfect ways. This brings only discord, for it is impossible to go left and right at the same time. And we must all go only one way together, to find GOD. If any soul thinks they have found an allowed unaccountable middle ground, they are truly insane.

Our firm desire and reliance towards GOD must be affirmed by HOLYSPIRIT; but it is our first firm enduring choice of direction that must be made. In this we see GOD'S purposed reasoned Principles are omnipotent and incorporate all who truly love GOD'S ways and truth. There is FATHER GOD, there is FATHER'S SON, and FATHER'S HOLYSPIRIT, which form the Trinity. Below the three facets and souls of GOD above all, we see the orders of all the remaining Holy Angels. All being one in GOD'S love, belief, understanding, and purpose. Below these, the Holy Saints and martyrs of mankind, being also one with GOD in all ways. These all are as GOD together, all experiencing and living in the purposed principled knowing ways and beliefs of FATHER, all having heard and learned their own immutable measure of loving omniscient understanding of belief, coming from FATHER alone. Again, in it all being other-worldly, this can only be understood by experiencing it for one's self. Acts 17:26-29, "*AND HATH MADE OF ONE BLOOD ALL NATIONS OF MEN FOR TO DWELL ON ALL THE FACE OF THE EARTH, AND HATH DETERMINED THE TIMES BEFORE APPOINTED, AND THE BOUNDS OF THEIR INHABITANTS; THAT THEY SHOULD SEEK THE LORD,*

IF HAPLY THEY MIGHT FEEL AFTER HIM, THOUGH HE BE NOT FAR FROM EVERY ONE OF US: FOR IN HIM WE LIVE, AND MOVE, AND HAVE OUR BEING; AS CERTAIN ALSO OF YOUR OWN POETS HAVE SAID, FOR WE ARE ALSO HIS OFFSPRING. FORASMUCH THEN AS WE ARE THE OFFSPRING OF GOD…" We are all gifted, created, and come from FATHER, to aspire and grow to be eternal children of GOD. Each true child, is known as FATHER is known. GOD'S true children are seen, as FATHER is seen. All true children exist as FATHER exists. FATHER'S true children know as he knows, and we love as he loves. Love is the most intimate knowing force there is; so, to in any way and measure, love someone, we must understand, know, agree with, and be bonded to them in all our foundational ways. And to love GOD *eternally*, means to completely understand, know, agree with, and be bonded with them in every gifted created lively way. It is known, that only after fully seeing and truly understanding exactly what Heaven is, and means, and knowing its eternal purpose, that we can know and truly understand what is required for us to live there together in love and peace for eternity. We all see that when a soul's love for another dies;

they say things like "I don't know you anymore" or "I can't understand you anymore" or "I don't agree with you anymore" this is a universal constant of love. So, if you don't understand, know, or agree with GOD in any way, your love for GOD is not its fullest measure. GOD'S true children will never surrender the defining parameters of this spiritual battle field to the fallen. Yet most weak professed believers, have allowed the Bible's parameters and definitions to be interpreted and set by the lost. GOD through us, defines this battlefield, not the lost. To see reality clearly, we must stand completely outside of it with GOD. So, not the possessive term "GOD'S", but the plural term "GODS" was spoken by FATHER in that he has truly said "ye are GODS" not ye are GOD'S. And "you are Children of the Most-High" Psalms 82:1-8, *"GOD STANDETH IN THE CONGREGATION OF THE MIGHTY; HE JUDGETH AMONG THE GODS. HOW LONG WILL YE JUDGE UNJUSTLY, AND ACCEPT THE PERSONS OF THE WICKED? SELAH. DEFEND THE POOR AND FATHERLESS: DO JUSTICE TO THE AFFLICTED AND NEEDY. DELIVER THE POOR AND NEEDY: RID THEM OUT OF THE HAND OF THE WICKED. THEY KNOW NOT, NEITHER WILL THEY*

UNDERSTAND; THEY WALK ON IN DARKNESS: ALL THE FOUNDATIONS OF THE EARTH ARE OUT OF COURSE. I HAVE SAID, YE ARE GODS; AND ALL OF YOU ARE CHILDREN OF THE MOST HIGH. BUT YE SHALL DIE LIKE MEN, AND FALL LIKE ONE OF THE PRINCES. ARISE, O GOD, JUDGE THE EARTH: FOR THOU SHALT INHERIT ALL NATIONS". The word "GOD*S*" minus an apostrophe, means *we each* are minute creative Gods. We are each a certain measure, gifted out from FATHER'S own forceful creative soul, to allow us our measured created awareness of reality; FATHER then having watched and known mankind, says in that last verse, but most die like animals, being a separated fallen physical mankind, rebelling and denying FATHER'S true created ways and desired eternal beliefs. Throughout Scripture, the word poor, most often refers to those spiritually poor in wisdom and truth, and not physical poverty. And "physically poor" in Scripture, means not having enough of life's required sustenance to physically maintain one's ability to work and live, not having a certain lesser level of provided standard of comfort and life. America's lowest financial class, would never be defined as poor in Scripture; they would be defined as physically blessed. Again, in John 10:33-38, *"THE JEWS ANSWERED HIM, SAYING, FOR A GOOD WORK WE STONE THEE NOT; BUT FOR BLASPHEMY; AND BECAUSE THAT THOU, BEING A MAN, MAKEST THYSELF GOD. JESUS ANSWERED THEM, IS IT NOT WRITTEN IN YOUR LAW, I SAID, YE ARE GODS? IF HE CALLED THEM GODS, UNTO WHOM THE WORD OF GOD CAME, AND THE SCRIPTURE CANNOT BE BROKEN; SAY YE OF HIM, WHOM THE FATHER HATH SANCTIFIED, AND SENT INTO THE WORLD, THOU BLASPHEMEST; BECAUSE I SAID, I AM THE SON OF GOD? IF I DO NOT THE WORKS OF MY FATHER, BELIEVE ME NOT. BUT IF I DO, THOUGH YE BELIEVE NOT ME, BELIEVE THE WORKS: THAT YE MAY KNOW, AND BELIEVE, THAT THE FATHER IS IN ME, AND I IN HIM"*. This last answer from JESUS, was worded to place their eternal accountability, fully and firmly upon their own shoulders. We must know and understand the many positive or negative effects we all bring upon reality, as forceful creative souls. Affects and effects brought by our projected patterns of forceful beliefs and understanding are all being acted out in this world. Each soul's every thought has its projected forceful measure of love or hate for GOD, for life or death, for oneness or individuality, for sustaining reality or destroying it all. The current circle of physical life and reality, presents itself as being in its final ongoing fall. Each generation inherits the greater failed legacy of the prior generation's effort and love, poured into their lives to try and define what spiritual life means and has become to them. Yet this fact is not recognized, understood, or appreciated, in any relatable way. And so, each generation proceeds to accuse and tear at the former, and build their own new desired self-righteous even-more inferior spirituality, racing towards mankind's final state, defined in Sodom and Gomorrah as being all zombie *militant* perversions. Again, please see how in a world of endless beliefs, good decisions for all people at large are impossible. Each decision leaders make, helps or hurts all individuals living under that decision to their own differing degrees, all as the individuals living under that decision, agree or disagree to differing degrees. So, no pure good choice can exist among endless beliefs. Why are there no perfect societies? Because there are so very, very few perfect people. Yet so many lost confused insane souls see themselves as good, and fighting against the bad. Of course the bad they see is personal, and born of their false individual self-defined good and evil. Only in a created perfect world where all souls share the one true existential belief and love given directly from GOD, can there ever exist perfect eternal decisions, and a perfect way of life for many. Until we love GOD with all we are, and seek out and know GOD'S ways and beliefs in our fullest adequate measure, we cannot be sure of one single understanding or desire we hold. But in *our* full and adequate measure of GOD, rests all our hope, wonder, and peace. We love and thank you HOLYSPIRIT, for the breath-taking wonder and beauty you are, and all you endure to unite us in your eternal truth; presenting yourself as loving, teacher, comforter, and Spiritual unifier, between all Godly souls. We love and thank you, JESUS, for your strength, beauty, and glory, of being who and all you are. For your unimaginable love and passion, you have endured in your hope for us to return our love to you. Who are we, that you would endure such noble brave things for us. You are our loving eternal KING, and we are your loving eternal subjects. We love and thank you, FATHER, for desiring an eternal loving family, which brought all the indeterminable breadths, depths, and heights, of your designed, reasoned, purposed, and created reality; and all your shared eternal loving goodness. You hold and share the one pure truth of life with all who desire to truly know, love, and live. GOD, we worship you above all that exists... FATHER GOD, you are seen by most as mystical and magical, yet you are real, eternal, and entirely spiritually, forcefully, and physically, defined to all who will and can see and hear you. And to all who are, or will be of GOD, you and I may also in certain ways seem mystical and magical; yet we too are real, eternal, and entirely spiritually, forcefully, and physically, defined before GOD. To better understand the forceful *creative* eternal powers of every soul, and our created physical substantive perceptions in a greater scientific and spiritual fullness. If you have not already, please read our other three books in order, listed on page two of this book, and available through me or on amazon. If in truth to GOD, these books costs are beyond your means. To receive a reduced-price copy, or to share GOD'S love concerning these or other spiritual things, then please email us as soon as you can at livetoknowtheGODofprophecy@gmail.com

Made in the USA
Monee, IL
04 November 2026

67383324R00174